Lecture Notes in Bioinformatics 13483

Subseries of Lecture Notes in Computer Science

More information about this subseries at https://link.springer.com/bookseries/5381

Davide Chicco · Angelo Facchiano ·
Erica Tavazzi · Enrico Longato ·
Martina Vettoretti · Anna Bernasconi ·
Simone Avesani · Paolo Cazzaniga (Eds.)

Computational Intelligence Methods for Bioinformatics and Biostatistics

17th International Meeting, CIBB 2021
Virtual Event, November 15–17, 2021
Revised Selected Papers

Springer

Editors
Davide Chicco (iD)
University of Toronto
Toronto, ON, Canada

Erica Tavazzi (iD)
Università di Padova
Padua, Italy

Martina Vettoretti (iD)
Università di Padova
Padua, Italy

Simone Avesani (iD)
Università di Verona
Verona, Italy

Angelo Facchiano (iD)
Consiglio Nazionale delle Ricerche
Avellino, Italy

Enrico Longato (iD)
Università di Padova
Padua, Italy

Anna Bernasconi (iD)
Politecnico di Milano
Milan, Italy

Paolo Cazzaniga (iD)
Università di Bergamo
Bergamo, Italy

ISSN 0302-9743   ISSN 1611-3349 (electronic)
Lecture Notes in Bioinformatics
ISBN 978-3-031-20836-2   ISBN 978-3-031-20837-9 (eBook)
https://doi.org/10.1007/978-3-031-20837-9

LNCS Sublibrary: SL8 – Bioinformatics

This Springer imprint is published by the registered company Springer Nature Switzerland AG
The registered company address is: Gewerbestrasse 11, 6330 Cham, Switzerland

Preface

This volume contains selected scientific papers of the 17th edition of the International Meeting on Computational Intelligence Methods for Bioinformatics and Biostatistics (CIBB 2021), which was the first online edition since its foundation. This scientific conference gathered scientists, researchers, scholars, and students working on computational intelligence, bioinformatics, biostatistics, and medical informatics from all over the world. More than in the previous editions, the virtual format of this CIBB 2021 allowed the participation of attendees from all habited continents of the planet, creating a unique transdisciplinary and international forum to discuss new challenges and trends in computational biology, biostatistics, and health informatics.

The conference program included only oral presentations, among which cutting-edge plenary keynote lectures were given by five preeminent keynote scientific speakers: Karsten Borgwardt (ETH Zürich, Switzerland) who talked about machine learning applied to data of patients in the intensive care unit, Ombretta Melaiu (Ospedale Pediatrico Bambino Gesù, Italy) who talked about application of computational statistics and machine learning to bioinformatics data of patients with neuroblastoma, André M. Carrington (Ottawa Hospital and Region Imaging Associates, Canada) who presented a new partial ROC AUC measure, Jacques Balayla (McGill University, Canada) who introduced the prevalence threshold metric, and Dmytro Fishman (Tartu Ülikool, Estonia) who presented the DOME recommendations for validation of machine learning results.

Moreover, our conference included two invited talks regarding themes functional to scientific research: Stefano Tonzani (lead editor at iScience, Cell Press, USA) presented some current editorial trends and challenges of modern scientific journals, while Olaf Wolkenhauer (Universität Rostock, Germany) gave a talk about successful grant writing. We organized the conference program with a main conference track, including short papers dealing with heterogeneous open problems at the forefront of current research, and four special sessions on specific themes: Artificial Intelligence and Statistical Methods for Neurodegenerative Diseases (in collaboration with the Horizon 2020 BRAINTEASER project), Modeling and Simulation Methods for Computational Biology and Systems Medicine, Towards Standardizing Machine Learning in Life Sciences: the FAIR Principles and the DOME Recommendations (in collaboration with the ELIXIR Europe Machine Learning Focus Group), and Machine Learning in Healthcare Informatics and Medical Biology. More information about the program is available on the conference website at http://www.isa.cnr.it/cibb2021/ or https://davidechicco.git hub.io/cibb2021/index.html.

The organization of the CIBB 2021 conference was supported by the IEEE Italy Section, the Italian chapter of the IEEE Computational Intelligence Society, and the Italian chapter of the IEEE Systems, Man, and Cybernetics Society.

The conference was held online on the WeConf.eu platform, during November 15–17, 2021, thanks to the synergistic effort of the general chairs, the advisory board, the Program Committee, and the technical chairs. The conference was attended by approximately 150 unique participants per session on average.

Following the principles for open science, we invited the authors of all short papers to release their manuscripts as preprints on preprint web servers (bioRxiv, medRxiv, and arXiv), to publish their software code on software code repositories (GitHub, GitLab, SourceForge, or others), and to openly release their datasets online on public data repositories (FigShare, Zenodo, Kaggle, University of California Irvine Machine Learning Repository, or others), to make their computational analyses reproducible and their data reusable.

For authorship, we invited the CIBB authors to follow the authorship guidelines of the International Committee of Medical Journal Editors (ICMJE), as recommended by the University of Toronto Faculty of Medicine.

A total of 68 short papers were submitted for consideration to CIBB 2021 and, after a round of reviews handled by the members of the Program Committee, 51 short papers were accepted for an oral presentation. After the conference, the authors of all accepted short papers were invited to submit an extended version of their manuscript to a supplement of two scientific journals (BMC Bioinformatics or BMC Medical Informatics and Decision Making) or to this Springer LNBI volume. This volume received 26 submitted chapters, and each chapter received three independent reviews; eventually 19 chapters were accepted for publication.

The editors warmly thank the conference participants, the authors, the keynote speakers, the Program Committee members, the advisory board delegate Roberto Tagliaferri, the reviewers, the CIBB 2019 general chair Paolo Cazzaniga, the special sessions' organizers, and anyone involved for their contributions to the success of CIBB 2021. The editors would like to send a special thank you to the general chairs and organizers of all the previous CIBB conference editions, from CIBB 2004 to CIBB 2019: their hard work, devotion, and patience built the foundations for CIBB 2021 and for all the future conference editions ("If I have seen further it is by standing on the shoulders of giants", Isaac Newton).

September 2022

Davide Chicco
Angelo Facchiano

Organization

General Chairs

Davide Chicco   University of Toronto, Canada
Angelo Facchiano   Consiglio Nazionale delle Ricerche, Italy
Margherita Mutarelli   Consiglio Nazionale delle Ricerche, Italy

Program Committee Chairs

Davide Chicco   University of Toronto, Canada
Angelo Facchiano   Consiglio Nazionale delle Ricerche, Italy
Enrico Longato   Università di Padova, Italy
Erica Tavazzi   Università di Padova, Italy
Martina Vettoretti   Università di Padova, Italy
Simone Avesani   Università di Verona, Italy
Anna Bernasconi   Politecnico di Milano, Italy
Paolo Cazzaniga   Università di Bergamo, Italy

Program Committee

Abbas Alameer   Kuwait University, Kuwait
Alessandro Guazzo   Università di Padova, Italy
Alessia Colonna   Università di Padova, Italy
Andrea Tangherloni   Università di Bergamo, Italy
Arnout Van Messem   Université de Liège, Belgium
Bayu Adhi Tama   University of Maryland, USA
Benjamin Huremagic   Katholieke Universiteit Leuven, Belgium
Binh Nguyen   Victoria University of Wellington, New Zealand
Burra Venkata Durga Kumar   Xiamen University, Malaysia
Chiara Roversi   Università di Padova, Italy
Christoph Friedrich   Fachhochschule Dortmund, Germany
Claudia Mengoni   Università di Verona, Italy
Daniele Maria Papetti   Università di Milano-Bicocca, Italy
Eleonora Auletta   Universidade de Lisboa, Portugal
Emanuel Weitschek   UniNettuno, Italy
Eugenio Del Prete   Istituto Telethon di Genetica e Medicina, Italy
Eva Viesi   Università di Verona, Italy
Francesca Marturano   Istituto Oncologico Veneto, Italy

Gabriel Cerono	University of California, San Francisco, USA
Giacomo Baruzzo	Università di Padova, Italy
Giovanni Birolo	Università di Torino, Italy
Giuseppe Agapito	Università Magna Graecia di Catanzaro, Italy
Hugo López Fernández	Universidade de Vigo, Spain
Isotta Trescato	Università di Padova, Italy
Iulian Ciocoiu	Universitatea Tehnică Gheorghe Asachi din Iași, Romania
Karin Verspoor	RMIT University, Australia
Laura Veschetti	Università di Verona, Italy
Marcio Dorn	Universidade Federal do Rio Grande do Sul, Brazil
Marco Nobile	Università Ca' Foscari Venezia, Italy
Marco Pansera	Consiglio Nazionale delle Ricerche, Italy
Marzio Pennisi	Università del Piemonte Orientale, Italy
Mikele Milia	Università di Padova, Italy
Mirko Treccani	Università di Verona, Italy
Pietro Bosoni	Università di Pavia, Italy
Roberto Di Marco	Università di Verona, Italy
Simone Riva	University of Oxford, UK
Simone Spolaor	Technische Universiteit Eindhoven, The Netherlands
Stefano Maestri	Università di Camerino, Italy
Tatiana Latychevskaia	Paul Scherrer Institute, Switzerland
Umberto Perron	Human Technopole, Italy
Vincenzo Bonnici	Università di Parma, Italy
William Yuan	Harvard Medical School, USA
Youngro Lee	Seoul National University, South Korea

Reviewers

Alessandro Palombit	Corsmed, Sweden
Alessandro Stefanini	Università di Pisa, Italy
Alessia Paglialonga	Consiglio Nazionale delle Ricerche, Italy
Alex Graudenzi	Consiglio Nazionale delle Ricerche, Italy
Alfredo Vellido	Universitat Politècnica de Catalunya, Spain
Andrea Bracciali	University of Stirling, UK
Andreia S. Martins	Universidade de Lisboa, Portugal
Andrey Rzhetsky	University of Chicago, USA
Angelo Ciaramella	Università di Napoli Parthenope, Italy
Angelo Facchiano	Consiglio Nazionale delle Ricerche, Italy
Anjany Sekuboyina	Technische Universität München, Germany
Anna Bernasconi	Politecnico di Milano, Italy

Sean Holden	University of Cambridge, UK
Stefano Rovetta	Università di Genova, Italy
Tiago Azevedo	University of Cambridge, UK
Tianlin Zhang	University of Manchester, UK
Tiziana Sanavia	Università di Torino, Italy
Umberto Ferraro Petrillo	Sapienza Università di Roma, Italy
Vassilis Plagianakos	University of Thessaly, Greece
Veronica Vinciotti	Università di Trento, Italy
Wellington Pinheiro dos Santos	Universidade Federal de Pernambuco, Brazil
Yair Goldberg	Technion Israel Institute of Technology, Israel
Yanqing Zhang	Georgia State University, USA

Technical Chairs

Angelo Facchiano	Consiglio Nazionale delle Ricerche, Italy
Antonella Iuliano	Università della Basilicata, Italy
Antonino Staiano	Università di Napoli Parthenope, Italy
Claudio Angione	Teesside University, UK
Davide Chicco	University of Toronto, Canada
Enrico Longato	Università di Padova, Italy
Erica Tavazzi	Università di Padova, Italy
Fotis E. Psomopoulos	Centre for Research and Technology Hellas, Greece
Giuseppe Agapito	Università Magna Graecia di Catanzaro, Italy
Manuel Tognon	Università di Verona, Italy
Margherita Mutarelli	Consiglio Nazionale delle Ricerche, Italy
Martina Vettoretti	Università di Padova, Italy
Simone Avesani	Università di Verona, Italy

CIBB Steering Committee (Advisory Board) Delegate

Roberto Tagliaferri	Università di Salerno, Italy

Webmasters

Davide Chicco	University of Toronto, Canada
Joao Ribeiro Pinto	INESC TEC and Universidade do Porto, Portugal

CIBB Steering Committee (Advisory Board)

Roberto Tagliaferri	Università di Salerno, Italy
Davide Chicco	University of Toronto, Canada
Francesco Masulli	Università di Genova, Italy

Elia Biganzoli	Università Statale di Milano, Italy
Clelia Di Serio	Università Vita-Salute San Raffaele, Italy
Pierre Baldi	University of California, Irvine, USA
Alexandru Floares	Solutions of Artificial Intelligence Applications Institute, Romania
Jon Garibaldi	University of Nottingham, UK
Nikola Kasabov	Auckland University of Technology, New Zealand
Leif Peterson	Methodist Hospital Research Institute, USA

CIBB Founders

Roberto Tagliaferri	Università di Salerno, Italy
Francesco Masulli	Università di Genova, Italy
Antonina Starita (1939–2008)	Università di Pisa, Italy

CIBB Past Editions

CIBB 2019, Bergamo, Italy
CIBB 2018, Caparica, Portugal
CIBB 2017, Cagliari, Italy
CIBB 2016, Stirling, Scotland
CIBB 2015, Naples, Italy
CIBB 2014, Cambridge, England
CIBB 2013 (co-organized with PRIB 2013), Nice, France
CIBB 2012, Houston, Texas, USA
CIBB 2011, Gargnano, Italy
CIBB 2010, Palermo, Italy
CIBB 2009, Genoa, Italy
CIBB 2008, Vietri sul Mare, Italy
CIBB 2007 (within WILF 2007), Camogli, Italy
CIBB 2006 (within FLINS 2006), Genoa, Italy
CIBB 2005 (within WILF 2005), Crema, Italy
CIBB 2004 (within WIRN 2004), Perugia, Italy

Contents

Chemical Neural Networks and Synthetic Cell Biotechnology: Preludes to Chemical AI

Pasquale Stano(✉)(iD)

Department of Biological and Environmental Sciences and Technologies (DiSTeBA), University of Salento, 73100 Lecce, Italy
pasquale.stano@unisalento.it

Abstract. Synthetic Biology and Artificial Intelligence are two relevant fields in modern science. Together with Robotics, they have either practical scopes, or can be used for modeling organisms' features and behaviors. The recent Synthetic Biology advancements in the so-called "synthetic cells" area allow the construction of cell-like systems with non trivial complexity, paving the way to a novel direction: the realization of chemical artificial intelligence. One possible path foresees the "installation" of chemical versions of artificial intelligence devices in synthetic cells. In this article we present this new scenario, focusing on chemical mechanisms and systems that are topologically organized as neural networks, highlighting their possible role in synthetic cell biotechnology. Future directions, challenges and requirements, as well as epistemological interpretations are also briefly discussed.

Keywords: Synthetic Biology · Synthetic cells · Chemical computation · Neural networks · Embodiment

1 Can "Synthetic Cell" Biotechnology Become a Useful Platform for Chemical AI?

The "Sciences of the Artificial" [1–3] aim at constructing artifacts capable of behaving like biological systems, in order to gain new scientific knowledge by the synthetic method. The latter is based on the *construction* of models (artifacts) that reproduce biological organization and behavior. Fields as Artificial Intelligence (AI), Robotics, and Synthetic Biology (SB) represent three different approaches, respectively, in the software, hardware, and wetware domains. Prompted by recent advancements in SB, here we intend to highlight the possible (but probably not so near) developments in this new wetware domain of tools and strategies, which have been traditionally explored in AI.

Biologically inspired methods have literally revolutionized AI. Approaches such as neural networks (NNs), genetic algorithms, and membrane computing – just to mention a few names – are typical examples of bio-inspired computing. No doubts that these approaches have contributed to the impressive success

D. Chicco et al. (Eds.): CIBB 2021, LNBI 13483, pp. 1–12, 2022.
https://doi.org/10.1007/978-3-031-20837-9_1

of present-day computational methods in AI. On the other hand, the current advancements in SB have generated a completely new biotechnology based on the so-called "synthetic cells" (SCs, also known as "artificial cells", and sometimes "protocells"). Thanks to the efforts of an ever increasing number of practitioners [4,5], the experimental horizon on SCs paves the way to unprecedented approaches, possibly leading to SCs capable of AI-like bio/chemical computation. In other words, a new generation of SCs could be endowed with capabilities which can be defined "intelligent" to a similar extent we do with AI artifacts.[1]

These forms of bio/chemical computation can be of two types. It is known that the operations carried out by metabolic, genetic, sensorial, and intercellular communication networks in biological cells can be interpreted as a form of analogical computing. Therefore, SCs – just because they are endowed with reconstructed biochemical networks – are computing machines by definition. This first approach is based on a plain imitation of biological counterparts. The second approach derives, instead, from a sort of reversal process, whereby current AI tools and strategies (which in turn can be bioinspired or not) are implemented in SCs, by properly designing *ad hoc* artificial chemical computation devices. This second way would lead to wetware systems with AI-inspired functions or modules that could display unique features when compared to the silicon counterparts. However, the potential developments here described are probably far from an imminent experimental reach. Nevertheless we believe that the current discussion can prompt further attention and drive the field into exciting directions.

The emerging reciprocal influence between SB and AI constitutes an interesting case for inquiring novel aspects of computation, and perhaps it can help exploring in original way the generative mechanisms of life and cognition. While Artificial Life (ALIFE) practitioners have traditionally studied these problems by means of hardware and software approaches, the wetware arena is still at its infancy. We therefore believe that the theoretical, computational, biological, chemical, epistemological aspects of this scenario (in our opinion rather imminent) are completely open to discussion, categorization, and criticism. Moreover they are particularly open to brilliant and creative proposals. More specifically, we envision several positive contributions stemming from the development of chemical AI devices grafted into SCs. The ultimate goal will be embarking on the path towards rudimentary forms of intelligent/cognitive SCs [6–8].

2   Scientific Background - What Exactly are SCs?

Pioneer efforts carried out in the 1990s have made it clear that it is possible to build cell-like systems from scratch, i.e., via guided-assembly and/or self-assembly processes applied to bio-organic or artificial molecules [9–12]. SCs

---

[1] Because current SCs still lie at a far lower complexity level when compared with living organisms (even the simplest ones), by "intelligent" SCs we mean systems that most resemble *machines* rather than *organisms*. It seems appropriate, for the moment, referring to SC "intelligence" in this narrow sense. See also Sect. 2.1.

consist in self-bounded systems constituted by a determined set of interacting molecules encapsulated in a compartment with a semi-permeable membrane [13], as shown schematically in Fig. 1a. Liposome technology, in particular, plays an important role because typical SCs are made of liposomes filled with bio-organic molecules such as enzymes and nucleic acids. Moreover, the SC membrane can host membrane proteins, especially when it is made of phospholipids. Finally, SC sizes typically range from ca. 0.1 to 10 μm.

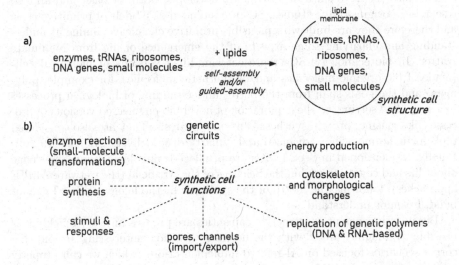

Fig. 1. Synthetic cell technology. (a) SCs can be constructed from separated molecules by techniques based on self- or guided-assembly. In particular, a common design starts from a selected set of biomolecules, such as enzymes, tRNAs, ribosomes, DNA genes (very often: plasmids), and other small molecule, which become co-encapsulated in a lipid vesicle. Many techniques, originated in liposome technology, are available. (b) A non-exhaustive list of reconstituted functions by means of SCs. It is important to mention that several studies report the success of reconstituting an individual function, while the integration of more function is still challenging. Nevertheless it is possible to imagine that SCs with complex behavior will be achieved in the near future.

SCs are built to mimic fundamental cellular functions, such as simplified forms of metabolism, DNA duplication, DNA transcription, protein synthesis, chemical signaling, and so on. The list of separately achieved functions is considerable, but commenting on them lies outside the scope of this article. It is sufficient to say that non-trivial systems can be constructed and that there is a consensus about the future expansion of this technology. However, while SCs share some similar – yet very simplified – features with biological cells, it is fair to say that current SCs are far from being alive, despite the above-mentioned advancements. For example, many SC functions have been constructed separately, and major efforts should be now devoted to their integration [14]. SCs are currently unable to sustain the whole-cell reaction network capable of producing *all* own components (membrane included): they cannot undertake the autopoietic (self-producing) dynamics [15], even if some aspects of homeostasis, growth, reproduction have been achieved, separately and in a very simplified form [16]. Moreover, most of current SCs are not energetically autonomous; in the large majority of studies, SCs operate like a "spring-toy". Their functioning depends on a pre-given high-energy molecular reservoir. Only very recently it has been shown that SCs can generate ATP via photo-phosphorylation [17,18].

It is also useful to clarify that SCs are developed both for basic and applied research. In origins-of-life studies, SCs are intended as models of primitive cells, and therefore they are built with plausible primitive chemicals, aiming at understanding and clarifying some aspects of the emergence of life from inanimate matter. In biotechnology, SCs represent simplified versions of biological cells (made of bio-organic molecules and/or synthetic molecules, for example, polymers) and the goals span from the better understanding of biological processes in a simplified setting, to the exploration of novel bio-engineered versions of processes like sensing, protein synthesis, enzyme catalysis. They are also considered tools for nano-medicine (advanced and "smart" drug delivery systems [19,20]). Finally, as mentioned in Sect. 1, SCs are privileged tools for gaining knowledge about life and cognition within the Sciences of the Artificial (the wetware ALIFE approaches), following the motto of the synthetic methodology: 'What I cannot build, I cannot understand'.

Because the current constructive capacities are progressing very quickly it is possible to imagine, albeit with the necessary caution, interesting around-the-corner scenarios focused on AI-related implementations, which we can properly recognize as chemical AI approaches. Before moving to Sect. 3, however, it is necessary a clarification related to the description of what SCs, intended as organized dynamical systems, are. Such a specification is needed because it mirrors our attitudes towards the way we understand operations in SCs, in particular with respect to the subject of *information* – a pivotal concept in AI.

2.1   Computer Gestalt [21] *vs.* Autopoiesis & Autonomy

One of the theories that most influences the research on SC is autopoiesis: a systemic theory, introduced by H. Maturana and F. Varela in the 1970s, which provides an operational definition of any living systems, describing them as "autopoietic machines", and equating life to cognition [15]. As mentioned, current SCs are not autopoietic, thus not alive. Very rarely it has been reported about SCs displaying forms of minimal autonomy, circular organization, and self-regulative processes. Understanding the still-missing features of SCs is indeed crucial for a correct epistemological framing of what current SCs actually are, and how typical AI themes such as the concept of information, its manipulation and communication, can be dealt with. The interpretation of SC operations within the information/communication domains in the machine or "computer Gestalt" perspective [21] (Fig. 2a) is indeed very common in the literature of living systems and SCs as well. In the cases of SCs and in the possible implementations of chemical AI devices, we believe it is still acceptable, as it clashes only at a minor extent with more encompassing (but still not owned) aspects of living and cognitive systems, i.e., organizational closure, autonomy & autopoiesis, which endow dynamical systems with a mind-like character [21] (Fig. 2b). The situation, however, is not so simple. The embodied feature of chemical AI systems could allow, at least in principle, to a realization of organizationally closed networks (or parts of them) whereby (i) the generation of the network components, and (ii) the relations of reciprocal production between the network components, become *de facto* self-entailing.

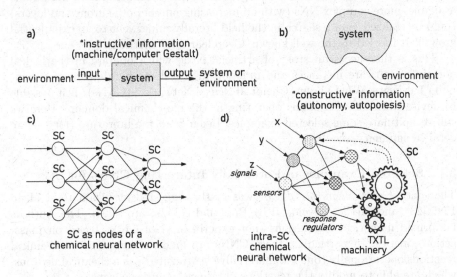

Fig. 2. Epistemic interpretations and chemical neural networks in SC research. (a) SCs seen from the perspective of "computer Gestalt", whereby information plays an instructive role. (b) SCs seen from the perspective of autonomy & autopoiesis, whereby SCs and the environment undergo structural coupling, and SC organization adapts – thanks to its plasticity – to environment. Information becomes co-constructive of the SC organization. (c) Chemical neural networks made of SCs interconnected to each other. Arrows indicate the exchange of chemicals, intended as input/output signals. (d) Chemical neural networks inside SCs. The shown example refers to two-components bacterial systems, made of membrane sensors (first layer) and response regulators (second layer). The latter play the role of regulators for gene activation, via TX-TL processes. Normal arrows indicate the transmission of information, in form of causal relations of transformation (i.e., phosphorylation). Dashed arrow indicate relations of production.

3   Bio-Chemical Neural Network

Artificial Neural Networks (NNs) are probably the most popular and successful approach in bioinspired AI. The history of NNs is well known [22]. In the 1940s

McCulloch and Pitts proposed that neurons could be modeled as simple logical circuits; in the 1950s Rosenblatt refined that model by creating a NN called *perceptron*, endowing neuronal inputs with a weight, and each artificial neuron with an activation threshold. However, it became clear that to replicate the highly interconnected structure of brain neurons, NNs need to be better designed as multiple layers (*multi-layer perceptron*), in a way that the outputs of first layer neurons feed the second layer neurons, and so on. Later on, research on NNs was somehow hindered and delayed due to the criticisms raised by Minsky in 1969. However, in the 1980s it raised again, also thanks to the introduction of back-propagation algorithms that facilitated the NN "training", and to other improvements. When, in the following decades, powerful computers were made available, more complex NNs (with an increasing number of neurons and layers) could be realized, and research in the field literally underwent to an exponential growth. It has led to the well known "deep learning" methods.

This is the successful story of artificial NNs, which operate in the logical domain of software instances and function in the electronic domain of a computer hardware. What about "Chemical Neural Networks" (CNNs)? Is it possible to devise wetware AI systems operating in the bio-chemical domain? Here we will recapitulate some selected examples, taken from the literature (for a more detailed summary, see [28]).

3.1   Selected Examples of Potentially Interesting CNNs for SCs

The concept of a *chemical "diode"* was firstly suggested by Okamoto et al. in 1987 [23], and further developed by Ross and collaborators [24]. The Authors performed numerical simulations (not experiments) of CNNs based on mass-action kinetics. In particular, these CNNs are made of hypothetical chemical reactions operationally (but not materially) "segregated" as a chemical neurons, which are able to modify the reactions of other chemical neurons.

Each i-th chemical neuron consists in a chemical network being in a steady state. The latter is achieved when some chemicals – marked with the asterisk in the scheme below – have constant concentrations (due to buffering or flowing). It results that the steady state concentrations of two key species in the network (A_i and B_i) strictly depend on the concentration of the catalyst C_i.

Reactions of the i-th chemical neuron:

$$I_{1i}^* \xrightleftharpoons{C_i} X_{1i}$$
$$X_{1i} + B_i \rightleftharpoons X_{2i}^* + A_i$$
$$X_{3i} + A_i \rightleftharpoons X_{4i}^* + B_i$$
$$X_{3i} \rightleftharpoons I_{2i}^*$$

The i-th chemical neuron can affect the dynamics of the j-th chemical neuron because A_i and B_i are allowed to activate or inhibit C_j. However, the j-th chemical neuron can be also affected by another (k-th) chemical neuron in similar way (C_j being activated or inhibited by A_k and B_k too, independently and additively with respect to the i-th neuron).

Ross and collaborators further explored the model to derive conditions for the construction of the logic gates AND, OR, j AND NOT k, NOR, as well as a chemical mechanism for synchronization in temporally discretized networks.

DNA *strand displacement* has been proposed as a convenient, predictable, and "universal" approach to molecular computing (universal because it can emulate the dynamics of any abstract chemical reaction network). DNA strand displacement experiments have been carried out. Single stranded DNA species can bind/unbind to DNA templates. Input strands are those that bind to the template, displacing the bound strands that function as output. Input and output strands can be connected to each other by means of a judicious choice of the templates and of the strands sequences, so that a DNA strand displacement network would emerges. Recent computational results have shown that these networks can become *adaptive* thanks to the so-called buffering strategy [25].

Theoretical work has demonstrated that synthetic gene networks are capable of *associative learning* [26], according to which simultaneous triggering of inputs strengthens the connection between those inputs, and – in this way – the system learns that they are associated. This has been realized in silico, by modelling a three-gene system regulated by three transcription factors and two repressors. Two input molecules can binds with the repressors, leading to loss of repression. The circuit "learns" to associate the two inputs together.

Interestingly, the same Authors have shown that the associative learning circuit can also be implemented via *phosphorylation networks* (see below); the basic requirement being the presence of proteins that can be single- or double-phosphorylated, such as MAPK protein kinases [26].

Kim et al. (2004) [27] proposed a biochemical model system that is a simplified analog of genetic regulatory circuits, based on the concept of the *transcriptional switch*. The model employs DNA templates, RNA polymerase to produce mRNA, and RNase to destroy mRNA. The network consists in a set of DNA templates and mRNA sequences interconnected via mRNA 'signals', which operate on the templates by switching them on/off. The study was based on simulations, and led to the conclusion that – with a few assumption – the biochemical network was mathematically equivalent to recurrent (Hopfield) neural networks. In this model, transcriptionally controlled DNA switches are treated as synapses, and the concentrations of mRNA species as the states of neurons.

An attempt to overcome the limitations of some models like those listed above (networks which are 'static' in their behavior, incapable of learning), Blount and collaborators devised a feedforward chemical neural network – based on chemical reaction networks – consisting of a network of cell-like compartments (in a nested topology), each containing a "chemical neuron" as a module. Some molecular species can permeate through the compartments boundaries, triggering the feedforward and back-propagation mechanisms [28]. The model is reminiscent of P *systems*, but reaction rates are modeled by realistic functions (mass action or Michaelis-Menten). The learning goal was focused on the binary XOR function, which was successfully implemented.

Finally, it is worthy of note the study by Hellingwerf and collaborators (1995) [29], who focused on phosphorylation pathways in *E. coli*, suggesting that *phosphorylation networks* actually resemble NNs in their connectivity. In particular, *two-component regulatory systems* were analyzed, highlighting the fact that although phosphoryl transfer pathways typically (but not always) take place in parallel, cross-talk is possible – in convergent and divergent manner – and it represents a useful mechanism for a NN design. The Authors maintain that these networks "may well lead to signal amplification, associative responses and memory effects, characteristics which are typical for neural networks". In our opinion, this sort of "phosphoneural networks" are approachable within SC technology.

In conclusion, we have reviewed, though not exhaustively, some possible CNNs for SC research. No claims have been made about the best design, although the "phosphoneural networks" seems quite resonant with current SC technology. Future numerical modeling and wet-lab experiments will be helpful in this respect.

4   Concepts and Experimental Perspectives on Chemical Neural Networks and Synthetic Cells

This paper is an opportunity to envision next-generation SCs as those structures endowed with a sort of minimal chemically embodied AI module(s), based on the implantation of CNNs (or "chemical perceptrons") in their dynamical organization. As a proof of concept, initial interest could be directed toward CNNs of *minimal complexity*. Consequently, the scenario we propose owes its interest and relevance mainly to the novelty of SC technology, and not to the performances of CNNs when compared to the ones working in the software domain. CNNs can be intended and designed as between-SCs (Fig. 2c) or within-SCs (Fig. 2d). While the first consists in high-order architectures whereby SCs themselves constitute the NN nodes (thus resembling the actual neuronal architecture) – exchanging chemicals as input/output signals, the latter refers to CNN made of reaction networks occurring inside SCs.

Here we will focus on CNNs inside SCs (Fig. 2d). Why are they important to progress SC technology? Let us present a tripartite discussion, focused on machine learning, meaning, embodiment.

4.1   Machine Learning

Artificial NNs are generally associated to the concept of *machine learning*. The goal of NN operations is pre-fixed by the designer (for example, recognizing a dog in a picture) and the 'learning' operation actually means the stepwise reduction of output errors, given a set of input data, by optimizing the neuron-neuron interaction strength and the activation threshold of each individual neuron. CNNs will be very different than NNs in machine learning. In particular, they will not be easily "trainable". Given a certain connectivity, the parameters of the CNNs (weights, thresholds) cannot be easily tuned. Ultimately these features depend upon intermolecular forces, molecular recognition, electrostatic and sterical fits. In turn, these physical factors are exquisite functions of 3D molecular structures and of the potential presence of effectors in the background (ions, solvent polarity, pH etc.). For example, when considered as a NN analogous, two-component regulatory systems in bacteria [29] have been submitted to a 'machine learning phase' (if we can call it in this way) during evolution, the resultant system being optimized to match certain environments and certain needed cellular responses, according to an adaptive dynamics. Is it possible to engineer two-component regulatory systems by at least *rewiring* the CNN connectivity? For example, the system could be rewired by directing the "output" of a first layer element (a kinase) toward a different response regulator in the second layer, rather than toward its own one (the cognate one). The modification of the interactions' strength is another option. Recent reviews are the good starting point for an up-to-date discussion about the technical possibilities, based on chimeric molecules, modular swapping, mutants [30,31]. A further mechanism could be based on time-delayed phosphorylation kinetics, which disfavors sensor-regulator specificity [32]. Similarly, the information flow in transcriptional NNs could be engineered by a proper reshuffling of promoters and genes.

4.2 Meaning

Referring again to two-component regulatory systems, it can be said that their function allows the integration of external and internal changes experienced by the dynamical system we call SC. When such a AI-device would be implanted in SCs, SC technology would progress because of the improved capability of information processing. A patter of signals (a 'situation') becomes the new 'input' signal, to be processed in order to lead to specific behavior (a pattern of gene expression). This perspective fits with the 'computer Gestalt' interpretation, or SCs as machines, and will mark a major step toward next-generation SCs, because SCs would be endowed with features that allow a mapping of environmental changes to specific internal patterns. This is equivalent to assigning (i.e., to design) meaning to certain situations. Accordingly, CNNs could play a role for the implementation of *semantic information* theories in SC research. We have already compared the application of Shannon- and semantic information theories in SC studies [33]. A first interesting direction that deserves further exploration refers to Donald M. MacKay operational definition of meaning (the selective function, exerted by a signal to a given receiver, on the set of possible transition probabilities between the states of readiness of an agent) [34]. A second direction is based instead on models and interpretations that highlight the co-dependent trajectories, expressed in probabilistic terms, of an agent in an environment, based on the internal organization of the first and on patterns that exist in the second [35].

4.3 Embodiment

But it is the profound difference between logical NNs and CNNs that has an additional, and more theoretical interest, because it contributes to model life and cognition according autopoiesis [15]. Even if we can analyze CNNs under the lens of Boolean logic and representations, their activity does not reside in the logical domain, because they do not operate on symbols, but on matter and structures. The CNN operations lie in the molecular (material) domain, a domain that also hosts the computing machine itself (the SC). Being molecular in their own nature, the CNN components can be in principle generated by the SC constitutive processes and thus *being themselves the product of their own computation.*[2] In other words, intra-SC CNNs would blur the classical distinction between instructions and data, or between computer and computed, and ultimately between mind and body – flooding into one of the most fascinating topic in the biology of cognition, self-referentiality, and second-order cybernetics [15]. Then, it would also incisively impact on wetware approaches to embodied AI, representing pioneering attempts to build chemical systems with features that are intrinsically not achievable in software and hardware implementations. For us, this would bring embodied AI one step closer to achieving artificial intelligent systems grounded on biological organization, moving beyond the mere behavioral imitations [7,8,36]

Acknowledgments. I am indebted to Luisa Damiano (IULM-Milan, Italy) and to Maurizio Magarini (Politecnico di Milano, Milan, Italy) for inspiring discussions.

References

1. Simon, H.A.: The Sciences of the Artificial. MIT Press, Cambridge MA (1996)
2. Cordeschi, R.: The Discovery of the Artificial. Behavior, Mind and Machines Before and Beyond Cybernetics. Springer, Netherlands (2002)
3. Damiano, L., Stano, P.: A wetware embodied AI? towards an autopoietic organizational approach grounded in synthetic biology'. Front. Bioeng. Biotech. **9**, 724023 (2021)
4. Schwille, P., et al.: MaxSynBio: avenues towards creating cells from the bottom up. Angew. Chem. Int. Ed. Engl. **57**, 13382–13392 (2018)
5. Frischmon, C., Sorenson, C., Winikoff, M., Adamala, K.P.: Build-a-Cell: Engineering a Synthetic Cell Community. Life **11**, 1176 (2021)
6. Cronin, L., et al.: The imitation game - a computational chemical approach to recognizing life'. Nature Biotech. **24**, 1203–1206 (2006)
7. Damiano, L., Stano, P.: Synthetic biology and artificial intelligence. grounding a cross-disciplinary approach to the synthetic exploration of (Embodied) cognition. Complex Syst. **27**, 199–228 (2018)

---

[2] Experimentally, it will not be easy to build SCs that *produce*, thanks to their internal metabolic processes, all components of the CNN. To start with, however, the SCs could be endowed with required components, according to the usual shortcut; or it could produce just a limited sub-set of CNN components.

8. Damiano, L., Stano, P.: On the 'Life-Likeness' of synthetic cells. Front. Bioeng. Biotech. **8**, 953 (2020)
9. Oberholzer, T., Wick, R., Luisi, P.L., Biebricher, C.K.: Enzymatic RNA replication in self- reproducing vesicles: an approach to a minimal cell. Biochem. Biophys. Res. Comm. **207**, 250–257 (1995)
10. Oberholzer, T., Albrizio, M., Luisi, P.L.: Polymerase chain reaction in liposomes. Chem. Biol. **2**, 677–682 (1995)
11. Oberholzer, T., Nierhaus, K.H., Luisi, P.L.: Protein expression in liposomes. Biochem. Biophys. Res. Comm. **261**, 238–241 (1999)
12. Szostak, J.W., Bartel, D.P., Luisi, P.L.: Synthesizing life. Nature **409**, 387–390 (2001)
13. Stano, P.: Is research on 'synthetic cells' moving to the next level? Life **9**, 3 (2019)
14. Abil, Z., Danelon, C.: Roadmap to Building a Cell: An Evolutionary Approach. Front. Bioeng. Biotech. **8**, 927 (2020)
15. Varela, F.J., Maturana, H., Uribe, R.: Autopoiesis: the organization of living systems, its characterization and a model. BioSystems **5**, 187–196 (1974)
16. Stano, P., Luisi, P.L.: Achievements and open questions in the self-reproduction of vesicles and synthetic minimal cells. Chem. Commun. **46**, 3639–3653 (2010)
17. Berhanu, S., Ueda, T., Kuruma, Y.: Artificial photosynthetic cell producing energy for protein synthesis. Nat. Commun. **10**, 1325 (2019)
18. Altamura, E., et al.: Chromatophores efficiently promote light-driven ATP synthesis and DNA transcription inside hybrid multicompartment artificial cells. Proc. Natl. Acad. Sci. USA **118**, e2012170118 (2021)
19. Chang, T.M.: Applications of artificial cells in medicine and biotechnology. Biomater. Artif. Cells Artif. Organs **15**, 1–20 (1987)
20. Leduc, P.R., et al.: Towards an in vivo biologically inspired nanofactory. Nat. Nanotechnol. **2**, 3–7 (2007)
21. Varela, F.J.: Principles of Biological Autonomy. Elsevier North Holland, New York (1979)
22. Wooldridge, M.: The Road to Conscious Machines. The Story of AI. Penguin Books, London (2020)
23. Okamoto, M., Sakai, T., Hayashi, K.: Switching mechanism of a cyclic enzyme system: role as a 'chemical diode'. BioSystems **21**, 1–11 (1987)
24. Hjelmfelt, A., Weinberger, E.D., Ross, J.: Chemical implementation of neural networks and Turing machines. Proc. Natl. Acad. Sci. USA **88**, 10983–10987 (1991)
25. Lakin, M.R., Stefanovic, D.: Supervised learning in adaptive DNA strand displacement networks. ACS Synth. Biol. **5**, 885–897 (2016)
26. Fernando, C.T., et al.: Molecular circuits for associative learning in single-celled organisms. J. R. Soc. Interface **6**, 463–469 (2009)
27. Kim, J., Hopfield, J., Winfree, E.: Neural network computation by in vitro transcriptional circuits. In: Saul, L., Weiss, Y., Bottou, L. (eds.) Advances in Neural Information Processing Systems 17 (NIPS 2004), pp. 681–688. Vancouver, Canada (2004)
28. Blount, D., Banda, P., Teuscher, C., Stefanovic, D.: Feedforward chemical neural network: an in silico chemical system that learns XOR. Artif. Life **23**, 295–317 (2017)
29. Hellingwerf, K.J., Postma, P.W., Tommassen, J., Westerhoff, H.V.: Signal transduction in bacteria: phospho-neural network(s) in Escherichia coli? FEMS Microbiol. Rev. **16**, 309–321 (1995)
30. Laub, M.T., Goulian, M.: Specificity in two-component signal transduction pathways. Annu. Rev. Genet. **41**, 121–145 (2007)

31. Agrawal, R., Sahoo, B.K., Saini, D.K.: Cross-talk and specificity in two-component signal transduction pathways. Future Microbiol. **11**, 685–697 (2016)
32. Skerker, J.M., Prasol, M.S., Perchuk, B.S., Biondi, E.G., Laub, M.T.: Two-component signal transduction pathways regulating growth and cell cycle progression in a bacterium: a system-level analysis. PLoS Biol. **3**, e334 (2005)
33. Magarini, M., Stano, P.: Synthetic cells engaged in molecular communication: an opportunity for modelling shannon- and semantic-information in the chemical domain. Front. Commun. Networks **2**, 48 (2021)
34. MacKay, D.M.: Information. Mechanism and Meaning. MIT Press, Cambridge MA (1969)
35. Kolchinsky, A., Wolpert, D.H.: Semantic information, autonomous agency and non-equilibrium statistical physics. Interface Focus **8**, 20180041 (2018)
36. Damiano, L., Stano, P.: a wetware embodied AI? towards an autopoietic organizational approach grounded in synthetic biology. Front Bioeng. Biotech. **9**, 873 (2021)

Development of Bayesian Network for Multiple Sclerosis Risk Factor Interaction Analysis

Morghan Hartmann[1]([✉]), Norman Fenton[1], and Ruth Dobson[2]

[1] School of Electronic Engineering and Computer Science, Queen Mary University of London,
London, UK
m.hartmann@qmul.ac.uk

[2] Wolfson Institute of Population Health, Queen Mary University of London, London, UK

Abstract. Extensive dataset availability for neurological disease, such as multiple sclerosis (MS), has led to new methods of risk assessment and disease course prediction, such as using machine learning and other statistical methods. However, many of these methods cannot properly capture complex relationships between variables that affect results of odds ratios unless independence between risk factors is assumed. This work addresses this limitation using a Bayesian network (BN) approach to MS risk assessment that incorporates data from UK Biobank with a counterfactual model, which includes causal knowledge of dependencies between variables. We present the results of more traditional Bayesian measurements such as necessity and sufficiency, along with odds ratios for each of the risk factors in the model. The greatest risk is produced by the genetic factor DRB15 (2.7 OR) but smoking, vitamin D levels, and childhood obesity may also play a role in MS development. Further data collection, especially in infectious mononucleosis in the population, is needed to provide a more accurate measure of risk.

Keywords: Bayesian networks · Risk assessment · Multiple sclerosis

1 Introduction

Multiple sclerosis (MS) is a neurological condition that causes lesions to form in the brain, brain stem, spinal cord, or optic nerve [1]. Since 1990, the prevalence of MS has increased by 10% with heightened numbers of cases reported in the United States, Canada, and Norwegian countries [2]. This general rise in MS cases, particularly in women, along with uncertainty around the pathogenic pathway and increasing availability of large datasets has increased research interest and efforts into the study of potential causes and triggers. Improved testing could account for the MS case increase, which further justifies using an approach that can model additional confounding variables.

Many medical decisions made under conditions of uncertainty can be modeled by creating a Bayesian Network (BN) [3–6]. Such medical application of BNs can involve condition monitoring, symptom manifestation, treatment effects, or diagnostics. A paper by Kyrimi et al. [7] focuses on creating BNs for different "idioms" or structures that are

© The Author(s), under exclusive license to Springer Nature Switzerland AG 2022
D. Chicco et al. (Eds.): CIBB 2021, LNBI 13483, pp. 13–24, 2022.
https://doi.org/10.1007/978-3-031-20837-9_2

common in medical reasoning. This paper focuses on a different idiom, the modeling of risk factors, which are exposures, habits, or demographic status of a patient that later increase the likelihood of developing a medical condition or disease, using MS as a specific example. Using a BN, interactions between variables are examined for their potential influence on MS development. Additionally, these variables often act as confounding influences, which can lead to some erroneously classified as risk-producing.

The goal of this analysis is to model genetic and environmental risk factors that could influence the risk of developing MS. To achieve this, an initial survey of risk factors and machine learning (ML) methods already used in MS research was conducted to identify variables to include in the network, which is discussed in Sect. 2. Section 2 also discusses the background of how BNs have previously been used an epidemiological tool. This leads to a larger explanation in Sect. 3 of how this BN was developed, including how the structure is determined and data integrated. This is extended to include an explanation of the measurements of necessity, sufficiency, and interaction, and how an odds ratio can be derived from BN observations. Finally, the results of the analysis are discussed in Sect. 4, where equivalent odds ratios found using the BN are compared to previous work.

2 Previous Work

2.1 Artificial Intelligence (AI) and Machine Learning (ML) in MS Research

A recent review was conducted on AI/ML methods that have been used in MS applications [8]. Most MS research using these methods involves lesion detection, prediction of clinical prognosis, and the analysis of immunological signatures. Recent work using MRI images as input to machine learning methods such as support vector machine, Naïve Bayes, and regression has been successful in classifying lesions versus brain matter. Similar methods have also been used to predict MS course type or future disability levels based on MRI scans and cognitive measures. Genetic risk factors have been studied using clustering, learning, and regression, but non-genetic factors have been neglected.

Bayesian methods have previously been used in a limited number of MS applications, including MS subtype prediction [9], treatment effects [10], and disease progression [11]. These papers all demonstrate the utility of Bayesian approaches, but do not include models that were derived from domain knowledge and logic. Overall, very little research using machine learning is used to analyze risk factors, and of those that do, none so far use Bayesian networks to show dependencies between variables.

2.2 Alignment with Epidemiology

Bayesian networks are not a standard tool for epidemiological studies, but by comparing the statistical concepts to epidemiological concepts and measures, it is possible to relate the two areas. The concept of directed causal diagrams for epidemiology was first proposed in 1999 by Greenland, Pearl, and Robins [12]. These diagrams, called directed acyclic graphs (DAGs), enable us to identify variables that contribute to confounding effects, but cannot represent interactions among variables. Based on Rothman's sufficient-component causes (SCC) framework, further work has been carried out to extend DAGs to represent sufficient cause structures [13]. However, any DAGs generated based on statistical equivalence do not necessarily represent causal relationships.

Traditionally, when attempting to find evidence for a causal relationship, epidemiologists look to the Bradford Hill (BH) criteria [14]. These rules require a proposed risk factor to have a temporal relationship with the disease, along with biological plausibility and other requirements based on cause and effect. A recent paper [15] highlighted the need for including these guidelines into causal diagrams and explains how DAGs are closely aligned with these. In a later section, this work will go through the criteria and show how these will be integrated into the BN model. Another paper [16] went further by combining the BH guidelines with graphical models (DAGs) to untangle the web of interactions amongst several exposures and genetic characteristics with Parkinson's disease. The authors go into great detail about possible DAG structures and analyse their plausibility within the guidelines. However, genetic confounding is different from exposure-related confounding and has a slightly different version of guidelines.

Bayesian Networks (BNs) are also a type of DAG, but they include probability tables that quantify the strength of dependencies between nodes. It is the goal of this work to show that, although BNs are not a standard tool for epidemiological studies, by comparing the statistical concepts to epidemiological concepts and measures, it is possible to relate the two areas. The publication upon which this work is based proposed a BN approach to modelling MS risk, and further explains the drawbacks of traditional epidemiological approaches and benefits of using BNs [8]. This work, authored by the same individuals of this paper, differs from [8] since UK Biobank data has been integrated into the model and epidemiological measures have been computed. Section 2 explains the structure of the BN model, the reasoning behind the encoded variable relationships, and how an odds ratio is derived from the counterfactual BN. Section 3 details the relative risk and odds ratio results obtained by running the model.

3 BN Development

A Bayesian network is an acyclic graph with an associated set of probability tables for each node [12]. A BN has three distinct components that make up the model—nodes and edges (which together define the graph structure) and node probability tables (NPTs). Nodes represent the variables included in the analysis, while the edges show direction of causal or statistical effect. Each node has an associated NPT which, for a node with parents, describes the probabilistic relationship between the node and its parents. In Fig. 1, the arrow (arc) from 'Smoking' to 'Multiple Sclerosis' represents a causal influence of smoking on MS risk. Before observations are entered, a Bayesian network contains only marginal probabilities based on the information entered in the NPTs and Bayesian inference methods. The basis for Bayesian networks is the ability to update prior probabilities based on evidence observed.

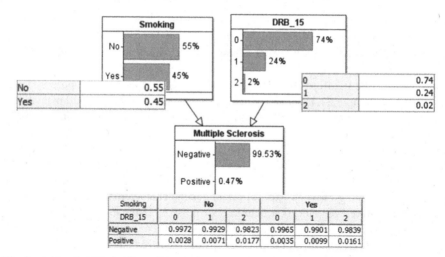

Fig. 1. A simple BN showing relationships between MS, Smoking, and Genetic Risk Factor DRB15 with associated probability tables

3.1   Relevant Risk Factors

The variables included in the model have been chosen since all have solid evidence as risk factors in MS literature. This is shown as a direct line from the risk factor to the 'MS' node in Fig. 2. See [17] for more detail on evidence for each risk factor presented. Ethnicity, Sex, and DRB15, a genetic factor in MS development, are the non-modifiable risk factors. In addition to genetic factors, it is believed that there are environmental triggers or influences that may affect the risk of developing MS. Risk factors included that can be modified to potentially alter the course of MS development are: infectious mononucleosis (IM), childhood body size, smoking, and vitamin D level. Some connections represent the non-biological connections that are due to confounding in the dataset. For example, cultural norms often influence whether a person smokes or how much vitamin D they receive through diet and sun exposure. Research by the Office of National Statistics published in 2019 revealed that the percentage of smokers among Asian (8.3%) and Black (9.7%) ethnic groups was lower than average compared to White (14.4%) and Mixed (19.5%) ethnic groups [18]. The same source also found that more men (15.9%) than women (12.5%) smoked. Due to these confounding factors, the NPTs for the Smoking and Vitamin D nodes depend on ethnicity and/or sex. The connection between ethnicity and the genetic risk factor, HLA DRB15, is based on biological evidence that individuals from White ethnic backgrounds are more likely to have this haplotype than non-White [19]. Guidance on monitoring vitamin D deficiency in London published by Barts and The London School of Medicine and Dentistry found that in a diverse east London population, a greater percentage of Black (47%) and South Asian (42%) individuals were deficient than White (17%) [20]. Based on these relationships, biological or not, are connected to each other and/or the MS node.

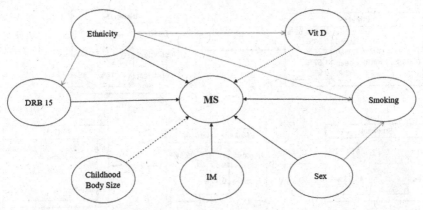

Fig. 2. Diagram of BN showing confounding influences (blue) and biological influences (black) (Color figure online)

3.2 Structure

Nodes can either be considered a direct risk factor for MS or a confounding factor that affects other risks. These specific factors were chosen based on a review of environmental and genetic factors contributing to MS risk [17] combined with data availability in UK Biobank. The relationships between these factors are outlined in Fig. 2. These connections that can be classified as biological in nature or have substantial literature backing are depicted in the black. If a relationship comes about through a non-biological effect, such as societal differences, the edge connection is in blue. There is only one black dotted connection, which signifies a weak association. However, it is included based on previous research findings and biological plausibility. The colour of the connection does not change any internal properties of the nodes; rather, it is simply for visualisation. The variables in gray represent the counterfactual world needed for the sufficiency and necessity measurements, which will be discussed further in the next section. NPTs were populated based on data directly from UK Biobank [21].

The data was filtered using Python scripts in the High Performance Compute cluster (HPC) supported by QMUL Research-IT. Figure 3 shows the result of training the BN on UK Biobank data. It is important to note that the MS node in this model is based on data on confirmed diagnoses of MS and will have some uncertainty due to unconfirmed cases. Similarly, the data used in model training for ethnicity, smoking, and childhood body size are self-reported and therefore might reflect recall bias. Figure 3 contains the marginal probabilities for each of the variables, which represented the percentages for the entire Biobank population when no characteristics are set to "True."

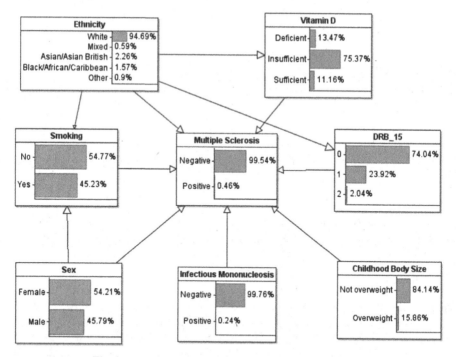

Fig. 3. Full BN structure containing marginal probabilities

3.3 Measurements

The next step is to estimate the effect of individual variables on the development of MS using the BN. To do this, we can merge basic concepts of epidemiological causation, such as sufficiency, necessity, and attribution, with Bayesian inference. By answering questions about the necessary and sufficient causes of MS, we can determine the most critical risk factors. This leads to an investigation into interaction, which occurs when the risk of developing a disease between individuals exposed and those not exposed to a given risk factor differs as a function of a third variable. For example, based on previous knowledge of MS risk factors, we would expect to see interaction between the genetic factor DRB15 and environmental risk factors.

Given the set of variables we have picked for MS risk, it can be shown through the BN that these are sufficient and/or necessary based on these definitions. However, these are both counterfactual questions, and therefore need a counterfactual model to address them. Since the UK Biobank data is an observational study, we can only perform this in a data analysis sense rather than a strict structural sense. The probability of sufficiency (PS) and the probability of necessity (PN) are defined by Judea Pearl in Causality [12]. The probability of sufficiency, which is the chance that any unexposed individual without the disease would have become affected had he or she been exposed can be represented in Eq. 1 below where d is the presence of disease, e is the exposure, while d' and e' are the absence of each, respectively. Similarly, Eq. 2 shows the probability of necessity. This refers to the chance that an exposed individual who developed a disease would have not developed the disease had he or she not been exposed. The counterfactual portion of each equation is contained in the first part of the probability. For example, d_e in Eq. 1 represents the counterfactual world probability that someone has the disease and were exposed to the risk factor given that in reality, they did not have the disease (d') and was not exposed (e').

$$PS \triangleq P\left(d_e | d', e'\right) \tag{1}$$

$$PN \triangleq P\left(d'_{d'} | d, e\right) \tag{2}$$

Odds Ratio is defined in Eq. 5 where $q+$ and $q-$ are the incidence rates in the exposed and unexposed populations, respectively. The equivalent expressions in probability terms that relate to this application are represented in Eqs. 3 and 4. The derived BN odds ratio (Eq. 5) is obtained by replacing $q+$ and $q-$ with their equivalent probabilistic expressions from Eqs. 3 and 4. Here the second term is considered the "bias" term; it represents the comparison between the probability that someone who is exposed does not have the disease and the probability that someone not exposed does not have the disease. In our study, the bias term is negligible since the prevalence of MS in the UK Biobank population is very low. This means that the probability odds ratio of developing MS approximates the relative risk.

$$q_+ \cong P(disease | exposed) = P(d|e) \tag{3}$$

$$q_- \cong P(disease | not exposed) = P(d|\neg e) \tag{4}$$

$$OR = \frac{q_+}{q_-} \times \frac{1 - q_-}{1 - q_+} \tag{5}$$

4 Results and Discussion

Using the measurements presented in Sect. 3.3, a clearer picture of how MS risk factors interact with each other and produce risk can be seen. This analysis starts with the Bayesian measures of sufficiency and necessity. These measurements can explain which risk factors are most influential in producing MS risk on their own and therefore offering insight into best targets for modification. Part 2 of this section includes the derived odds ratios and how they compare to previous work in the same dataset.

4.1 Interaction, Sufficiency, Necessity

In this section, the BN measurements of sufficiency and necessity (presented in Sect. 3.3) are assessed in this particular case. It might be argued that the risk factors included in the BN are, individually, very weak in terms of necessity or sufficiency. This is expected since MS is thought to be caused by a combination of factors rather than one single cause. It is important to note however that these weaker, more distal causes might be the most practical way to prevent MS development in terms of the UK population. To present an example of one such calculation, the haplotype DRB15 can be analyzed in terms of its necessity and sufficiency in producing MS on its own (shown in row 1 of Table 1 below). *PS* corresponds to the chance that any DRB15 negative individual without MS would have become affected had they instead been born with the DRB15 haplotype. *PN* value corresponds to the chance that an individual with DRB15 haplotype(s) who developed MS would have not developed MS had they not been born with this haplotype. Figure 4 shows a sample observation of how we obtain *PN* for the case DRB15, where the left-hand portion of the model shows the observations *d* and *e* needed for the calculation. The right-hand box shows the counterfactual world, which contains the observation *e'* and the resulting probability of necessity. There are some differences between the risk estimate obtained by the counterfactual BN and a typical measure such as relative risk or odds ratio. By using the BN, real-world observations can update probabilities of all unobserved risk factors (such as Smoking or Vitamin D status) which then are applied to the counterfactual world.

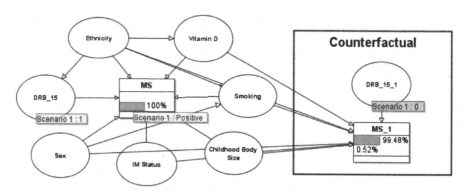

Fig. 4. BN with DRB15 counterfactual model

The concepts of sufficiency and necessity can be extended to Table 1 which shows the measures of necessity and sufficiency in developing MS. Since the risk of developing MS is very low to begin with, it can help to compare the sufficiency and necessity to the baseline risk of MS. The marginal MS probability ($P(MS)$) and its reciprocal ($P(notMS)$) are 0.46 and 99.54 percent, respectively. Therefore, from this dataset, we can conclude that each one of these variables are not at all necessary in developing MS. Some, like 2 DRB15 genes, may be sufficient on their own to produce MS. According to this model, none of these risk factors are necessary, and is most likely acting in conjunction with other component causes to promote MS.

Table 1. Sufficiency and necessity calculations for MS risk factors

Risk factor	PS (%)	Increase from marginal P(MS) (%)	PN (%)	Increase from marginal P(not MS) (%)
DRB15 (2 genes)	2.4	422.6	99.59	0.05
DRB15 (1 gene)	0.84	83.4	99.48	−0.06
Smoking	0.56	21.5	98.79	−0.75
IM	3.35	627.7	99.44	−0.097
Sex	0.64	38.1	99.15	−0.4
Childhood body size	0.55	19.3	98.69	−0.9
Vitamin D (deficient)	0.81	75.3	97.68	−1.9
Vitamin D (insufficient)	0.42	−7.9	98.67	−0.9

4.2 Equivalent Odds Ratios

The final step in exploring MS risk factor relationships is to compare the odds ratios the BN produces to the odds ratios produced by traditional epidemiological techniques. The source of odds ratios comes from a recently published UK Biobank study assessing gene-environment interactions [22]. We compare IM status, smoking, and adolescent body size odds ratios to the results obtained from the BN. The calculations for the odds ratios use values that come directly from the BN after observations are entered. There are some differences between the risk estimate obtained by the BN and a more typical measure such as relative risk or odds ratio. Using the result of $P(MS|x)$, which is the probability of MS given that a variable x is observed, taken directly from the BN, the equivalent relative risk and odds ratio can be calculated for each risk factor. In this case, x refers to each variable in column 1 in Table 2. This measurement differs from traditional epidemiological methods since it takes into account the dependencies between the risk factors. Tables 2 and 3 show the results of these two methods' calculations as well as a comparison between previous published results. No value for DRB15 is contained in the previous results since it was determined to be a confounding covariate.

4.2.1 Method 1

Method 1 uses each variable's relative risk (RR) and odds ratio (OR). The next step in verifying this relationship is to test whether the odds ratios the BN produces are approximately equal to the odds ratios produced by traditional epidemiological techniques. The source of odds ratios comes from a recently published UK Biobank study assessing gene-environment interactions [22]. We compare IM status, smoking, and adolescent body size odds ratios to confirm that similar results can be obtained from the BN. The calculations for the odds ratios use values that come directly from the BN after observations are entered. The large discrepancy in OR derived from the BN and through traditional means is similar in most cases, except for IM status. This could be due to extensive missing values in this field and over-reporting in MS cases. Table 2 summarizes these results.

Table 2. Odds ratio calculations using method 1

x	State	P(MS|x)	RR	OR (BN)	OR [22]
Smoking	True	.00527	1.291	1.293	1.21 (1.08–1.34)
	False	.00408			
Childhood body size	Plumper	.005538	1.243	1.244	1.36 (1.2–1.55)
	Normal/Thinner	.004455			
IM	Positive	.0335	7.351	7.571	1.82 (1.03–3.22)
	Negative	.004557			
DRB15	0	.003082	2.708	2.723	—
	1	.008349			

4.2.2 Method 2

A second way of determining an odds ratio from a BN is by using the AgenaRisk software to model each variable by itself outside of the relationships determined above. This would give an odds ratio that assumes independence with other variables. Two different models are needed to test the hypothesis of dependence between the proposed risk factor and MS development. The first model tests the likelihood that MS is dependent on the risk factor's presence. There are two estimates, one for the population with the risk factor and one for the population without. Once both likelihoods are found, the product of these is used as the likelihood estimate of the dependent hypothesis being true. The second model represents the scenario that the risk factor is independent from the outcome. The probability comparison for both hypotheses is found in columns three and four of Table 3. Through this analysis, we find that since the probability of the dependent hypothesis is

Table 3. Odds ratio calculations using method 2

r	State	Dependent	Independent	Risk ratio (Via BN)	OR ([22])
Smoking	True	**.0044**	.0027	0.922 (0.89–0.95)	1.21 (1.08–1.34)
	False				
Childhood body size	Plumper	**.0014**	.000019	0.762 (0.73–0.79)	1.36 (1.2–1.55)
	Normal/Thinner				
IM	Positive	**.0043**	.00201	0.385 (0.31–0.44)	1.82 (1.03–3.22)
	Negative				
DRB15	0	**.0022**	3.3E–7	1.43 (1.38–1.47)	—
	1				

greater in each case, it is assumed that each of the tested variables contribute to MS risk. The second part of this analysis is finding an estimate of the odds ratio from the BN.

From these results, it can be concluded that the clearest difference between exposed and non-exposed individuals is the genetic factor DRB15. This agrees with current Northern European studies, which have defined this factor as one of the most important genetic risk factors [17]. However, it is a possibility that smoking and childhood body size could also have a contribution to MS development according to this method. In literature, smoking has mixed results as a possible risk factor. Therefore, the result obtained here seems to agree with previous evidence as a weaker factor. The dependency between childhood body size and MS found in this paper agrees with a publication confirming an association between genetic predisposition to obesity and MS [23]. It is normal to see differences in the risk ratio and odds ratio since the risk ratio is the ratio of two probabilities whereas the odds ratio is the comparison of two events (e.g. MS versus no MS). Often the odds ratio will slightly overestimate the effect of an exposure. Method 1 provides a picture of MS risk in terms of the whole network of interacting factors, whereas Method 2 presents a way to determine whether a factor truly influences MS risk. The next step in this project will be validating with an external dataset and presenting the results.

5 Conclusions

By incorporating dependency relationships between variables, the BN adds an additional dimension to MS risk assessment. Inevitably, this results in différence between the risk measurements compared to previous estimates. A deeper look into the mechanisms within the BN that produce the large difference in OR for IM status is needed next, as well as a general look into how this measure can be translated into usable knowledge for decision making. This work can also be extended to a unique assessment of interaction between one or more risk factors.

References

1. Reich, D., Lucchinetti, C., Calabresi, P.: Multiple sclerosis. N. Engl. J. Med. **378**, 169–180 (2018)
2. G. 2. M. S. Collaborators: Global, regional, and national burden of multiple sclerosis 1990–2016: a systematic analysis for the Global Burden of Disease Study 2016. Lancet Neurol. **18**(3), 269–285 (2019)
3. Rose, C., Smaili, C., Charpillet, F.: A dynamic Bayesian network for handling uncertainty in a decision support system adapted to the monitoring of patients treated by hemodialysis. In: 17th IEEE International Conference on Tools with Artificial Intelligence (ICTAI 2005) (2005)
4. Jiang, X., Wells, A., Brufsky, A., Neapolitan, R.: A clinical decision support system learned from data to personalize treatment recommendations towards preventing breast cancer metastasis. Plos One **14**(3), 1–18
5. Fenton, N.E., Neil, M., Osman, M., McLachlan, S.: COVID-19 infection and death rates: the need to incorporate causal explanations for the data and avoid bias in testing. J. Risk Res. 1–4 (2020)

6. Neves, M.R., et al.: Causal dynamic Bayesian networks for the management of glucose control in gestational diabetes. In: 2021 IEEE 9th International Conference on Healthcare Informatics (ICHI), pp. 31–40 (2021). https://doi.org/10.1109/ICHI52183.2021.00018

7. Kyrimi, E., Neves, M., Neil, M., Marsh, W., McLachlan, S., Fenton, N.E.: Medical idioms for clinical Bayesian network development. Artif. Intell. Med. (2020)

8. Hartmann, M., Fenton, N., Dobson, R.: Current review and next steps in artificial intelligence in multiple sclerosis risk research. Comput. Biol. Med. **132** (2021)

9. Rodríguez, J., Pérez, A., Arteta, D., Tejedor, D., Lozano, J.: Using multidimensional Bayesian network classifiers to assist the treatment of multiple sclerosis. IEEE Trans. Syst. Man Cybern. Part C Appl. Rev. **42**(6), 1705–1715 (2012)

10. Pozzi, L., Schmidli, H., Ohlssen, D.I.: A Bayesian hierarchical surrogate outcome model for multiple sclerosis. Pharm. Stat. **15**, 341–348 (2016)

11. Bergamaschi, R., et al.: Immunomodulatory therapies delay disease progression in multiple sclerosis. Mult. Scler. J. **22**(13) (2016)

12. Pearl, J.: Causality, 2nd edn. Cambridge University Press, New York (2009)

13. VanderWeele, T.J., Robins, J.M.: Directed acyclic graphs, sufficient causes, and the properties of conditioning on a common effect. Am. J. Epidemiol. **166**(9) (2007)

14. Celentano, D.D., Szklo, M.: Gordis Epidemiology, 6th edn. Elsevier, Philadelphia (2019)

15. Shimonovich, M., Pearce, A., Thomson, H., Keyes, K., Katikireddi, S.V.: Assessing causality in epidemiology: revisiting Bradford Hill to incorporate developments in causal thinking. Eur. J. Epidemiol. **36**(9), 873–887 (2020). https://doi.org/10.1007/s10654-020-00703-7

16. Geneletti, S., Gallo, V., Porta, M., Khoury, M.J., Vineis, P.: Assessing causal relationships in genomics: from Bradford-Hill criteria to complex gene-environment interactions and directed acyclic graphs. Emerg. Themes Epidemiol. **8**(5) (2011)

17. Ramagopalan, S., Dobson, R., Meier, U.C., Giovannoni, G.: Multiple sclerosis: risk factors, prodromes, and potential causal pathways. Lancet Neurol. **9**, 727–739 (2010)

18. O. f. N. Statistics: Cigarette smoking among adults (2021)

19. Alves-Leon, S.V., Papais-Alvarenga, R., Magalhaes, M., Thuler, L.C., Fernandez, O.: Ethnicity-dependent association of HLA DRB1-DQA1-DQB1 alleles in Brazilian multiple sclerosis patients. Acta Neurol. Scand. **115**(5), 306–311 (2007)

20. Barts and The London School of Medicine and Dentistry Clinical Effectiveness Group. Vitamin D Guidance (2011)

21. UK Biobank: Protocol for a Large-Scale Prospective Epidemiological Resource (2007)

22. Jacobs, B.M., Noyce, A.J., Bestwick, J., Belete, D., Giovannoni, G., Dobson, R.: Gene-environment interactions in multiple sclerosis. Neurology: Neuroimmunol. Neuroinflamm. **8**(4) (2021)

23. Mokry, L.E., Ross, S., Timpson, N.J., Sawcer, S., Davey Smith, G., Richards, J.B.: Obesity and multiple sclerosis: a Mendelian randomization study. PLoS Med. **13**(6) (2016)

Real-Time Automatic Plankton Detection, Tracking and Classification on Raw Hologram

Romane Scherrer[1]([⊠]), Rodrigue Govan[1], Thomas Quiniou[1], Thierry Jauffrais[2], Hugues Lemonnier[2], Sophie Bonnet[3], and Nazha Selmaoui-Folcher[1]

[1] ISEA, Université de la Nouvelle-Calédonie, Nouméa, New Caledonia
romane.scherrer@hotmail.fr, nazha.selmaoui@unc.nc
[2] Ifremer, UMR9220 Entropie, Nouméa, New Caledonia
[3] Aix Marseille Univ, Université de Toulon, CNRS, IRD, MIO, Marseille, France

Abstract. Digital holography is an imaging process that encodes the 3D information of objects into a single intensity image. In recent years, this technology has been used to detect and count various microscopic objects and has been applied in submersible equipment to monitor the *in situ* distribution of plankton. To count and classify plankton, conventional methods require a holographic reconstruction step to decode the hologram before identifying the objects. However, this iterative and time-consuming step must be performed at each frame of a video, which makes it difficult to support real-time processing. We propose a real-time object detection based approach that simultaneously performs the detection, classification and counting of all plankton within videos of raw holograms. Experiments show that our pipeline based on YOLOv5 and SORT is fast (44 FPS) and can accurately detect and identify the plankton among 13 classes (97.6% mAP@0.5, 92% MOTA). Our method can be implemented to detect and count other microscopic objects in raw holograms.

Keywords: Object detection · Multiple Object Tracking · Deep learning · Plankton · Digital holography

1 Introduction

The observation, counts and classification of marine plankton are essential to measure the health of our oceans. In recent years, several submersible equipment [8] (ISIIS, LISST-Holo, eHoloCam) have been deployed as part of large-scale campaigns to acquire *in situ* images of plankton. Some of these systems use digital holography [14], a method that enable high resolution images acquisition over a large water column and at high flow rates. Since a hologram encodes the 3D information of all plankton as a single intensity image, a decoding process, called holographic reconstruction, is required to retrieve the sample image from its hologram. Unfortunately, the methods used to process holograms and then count and classify the species are still very time-consuming and manual.

D. Chicco et al. (Eds.): CIBB 2021, LNBI 13483, pp. 25–39, 2022.
https://doi.org/10.1007/978-3-031-20837-9_3

With the multiplication of collected images, various efforts have been made to accelerate and improve the holographic reconstruction, for instance, by adopting a convolutional neural network (CNN) to automatically find the focus [18] or to reconstruct a de-focused hologram without performing an auto-focusing or phase recovery routine [16,21]. Even though those approaches greatly accelerate the holographic reconstruction, the detection and classification of the objects need to be performed afterwards.

To count and identify the objects in a live video stream, three different tasks are necessary: (i) a classification task to identify the objects, (ii) a detection task to locate them and (iii) a tracking task to determine their respective trajectories to avoid counting the same objects several time during the video life span. However, these three distinct, yet complementary, tasks are often performed independently on holograms. The classification is often done on cropped holograms with, for example, a trained CNN as in [4,22] but a preliminary detection is necessary to determine those regions of interest (ROIs) that are then feed into the model. To detect the objects, some works have implemented a CNN-based sliding window algorithm [19] that perform a binary classification on different regions in the holograms to detect and count cells. Other studies propose to perform the detection with a segmentation-based method. The segmentation can be carried out with a threshold as in [17] that proposes to filter the intensity of the reconstructed holograms with a bandpass filter before applying a threshold to generate a binary mask. The segmentation can also be done with a deep learning model as in [7] where a Segnet model coupled with a circular Hough transform are applied on the holograms to locate the objects. However, detection by segmentation often requires a prior holographic reconstruction, as the diffraction patterns on raw holograms do not allow the object's boundaries to be precisely determined. Concerning the tracking task, which is performed to determine the objects trajectories, the existing methods are generally based on a frame-by-frame detection of the objects that are then associated through the sequences [10]. In the framework of holography, the detection assignment can be carried out with the calculation of the cross-correlation between two consecutive frames [13] which is effective when there is little variation in object morphology or noise between the images. When the motion of the objects causes a variation of their morphology (spin, rotation) between frames, other more robust algorithms, such as the minimum boundary filter (MBF) [9], have been successfully applied. However, these methods rely on a detection pipeline that requires a holographic reconstruction at each frame of the video.

Even if several approaches have been proposed in the last few years to detect and classify objects on holograms, the methods often focus on only one aspect, either a classification or a detection/tracking task. Moreover, most of the existing methods require a prior holographic reconstruction to detect the objects [20]. However, conventional algorithms [5] used to search each object's focus plane and remove twin image artifacts are iterative and computationally intensive and therefore not always compatible with real-time processing. Therefore, the use of an object detection model such as [12] Faster-RCNN, YOLO, SSD or RefineDet, offers an alternative by performing in real time the localization and classification of all objects on a frame in a single pass. Applied to raw holograms, these real-time models could greatly improve the applicability of digital holography and are compatible with other tracking algorithm to accurately count and classify the objects.

The aim of the paper is to demonstrate that the classification, localization and tracking of plankton can be simultaneously performed in real-time on raw holograms with an object detection model. For that purpose, two datasets of labeled in-line holograms will be simulated with 13 different plankton species. The paper is organized as follows. In the next section, the generation of holographic datasets and the object detection models are described. Section three shows the performances of the models. Conclusions and perspectives are given in the last section.

2   Materials and Methods

2.1   Hologram Formation

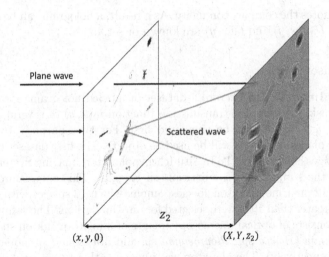

Fig. 1. In line-holography.

For an in-line holographic setup (see Fig. 1), the reference and object waves share the same optical axis and an object can be described by a complex transmission function [6] at a given z plane:

$$t_z(x,y) = \exp[-a(x,y)]\exp[i\phi(x,y)] \tag{1}$$

where $a(x,y)$ describes the absorption of the object and $\phi(x,y)$ is the phase distribution. The transmission function can be used to calculate the wavefront just behind the object $U_{z+}(x,y)$:

$$U_{z+}(x,y) = t_z(x,y)U_{z-}(x,y) \tag{2}$$

where $U_{z-}(x, y)$ is the incident wave that can be either plane or spherical.

Considering that the object is located at $z = 0$, the exit wave given by Eq. 2 can be rewritten as $U_{0+}(x, y) = t_0(x, u)U_{0-}(x, y)$ and is propagated to the detector/hologram plane which is located at $z = z_2$ along the optical axis. This propagation is simulated by the angular spectrum method by calculation of the following transformation:

$$U_{z_2}(X, Y) = TF^{-1}\left[TF\left(U_{0+}(x, y)\right) \times \exp\left(\frac{2\pi i z_2}{\lambda}\sqrt{1 - (\lambda u)^2 - (\lambda v)^2}\right)\right] \quad (3)$$

where λ is the wavelength and (u, v) are the Fourier domain coordinates. TF^{-1} and TF denoted the inverse and the direct Fourier transform, respectively. Note that Eq. 3 is often expressed as $U_{z_2}(X, Y) = R(X, Y) + O(X, Y)$ where R and O are the reference and the object waves that interfere at the surface of the recording medium. The recorded hologram at $z = z_2$ is the intensity calculated by:

$$H_{z_2}(X, Y) = |U_{z_2}(X, Y)|^2 = U_{z_2}(X, Y)U_{z_2}^*(X, Y) \quad (4)$$

where * denotes the complex conjugate. As a result, a hologram can be simulated once λ, z_2, $U_{0-}(x, y)$ and $t_0(x, y)$ are known or set.

2.2 Dataset

Plankton Images. To generate a dataset of labeled holograms for an object detection task, the complex transmission function $t_0(x, y)$ of several objects in a plane $(x, y, z = 0)$ must be simulated first. For that purpose, two labeled datasets of plankton images will be used as objects. The first dataset consists of shadow images collected by the In Situ Ichthyoplankton Imaging System (ISIIS), which was the subject of a competition on Kaggle[1]. This open source dataset consists of 121 marine plankton species, among which 10 species with a number of images greater than 1000 were selected for our simulations. The second dataset (custom) consists of optical microscopy images of 3 phyto-plankton species from New Caledonia (*Haslea sp.*, *Pleurosigma sp.* and *Mastogloia sp.* noted P1, P16 and P17, respectively). The plankton was imaged with a bright-field microscope at a ×10 magnification. The images were automatically thresholded, segmented into ROIs using an edge detection based algorithm (Sobel) and manually labeled. Figure 2 presents the number of images per species. Note that for each dataset, the ROI segments are labeled per class and saved as grayscale images. Moreover, the images were processed so that background has a constant value equal to 1 and only the pixels inside the object support have a value between 0 and 1. This particularly allow us to simulate the absorption $a(x, y)$ and the transmission function $t(x, y)$ of the objects. In particular, we converted a ROI segment $I(x, y)$ into an absorption with $a(x, y) = -1 \times I(x, y) + 1$ so that the transmission function is $t_0(x, y) = \exp[-a(x, y)]\exp[i\phi(x, y)]$ inside the object support and $t_0(x, y) = 1$ where there is no object $(a(x, y) = 0$ and $\phi(x, y) = 0)$. Note that

---

[1] https://www.kaggle.com/c/datasciencebowl/.

$t_0(x, y) = 1$ only implies that the incident wave that illuminates the sample remains undisturbed where there is no plankton ($U_0^+(x, y) = U_0^-(x, y)$).

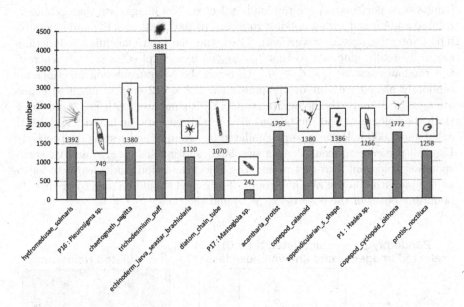

Fig. 2. Number of images per species.

To simulate $t_0(x, y)$ with various objects, the transmission functions of several plankton images can be randomly placed on a $N \times N$ empty (all-ones) image. By doing so, the (x, y)-axis coordinates of the bounding boxes are randomly set. Moreover, the plankton images are already saved as ROIs so that the bounding boxes width and height are the images dimensions. Since the images are classified per species, the labels of a simulated $t_0(x, y)$ for an object detection task (classes and bounding boxes coordinates) can be completely set. Once $t_0(x, y)$ is simulated, the corresponding hologram can be computed with the Eq. 3 and Eq. 4.

Holograms Simulation. To demonstrate that it is possible to classify and track objects on raw holograms in real time, we have simulated two datasets. The first dataset, used to train and test the detection model, consists of 10,000 simulated holograms. The second dataset, used to evaluate the tracking performance of the model, is composed of 100 simulated videos in which plankton are moving in a laminar flow in a two-dimensional plane channel. In this section, we describe in more detail the simulation of these two datasets.

Object Detection Dataset. Before simulating the transmission functions and the corresponding holograms to train the detection model, the plankton images from the two sample image datasets (ISIIS and Custom) were randomly split, per class, in a 80:20 ratio for training and testing, respectively. We have considered that the plankton are pure amplitude objects so that $\phi(x, y) = 0$. The simulation of $t_0(x, y)$ proceeds as follows. First, for each simulated $t_0(x, y)$, 13 plankton images (one per species) are randomly selected. The images are then randomly rotated and flipped with 4 possible rotations ($0°, 90°, 180°$ or $270°$) and 3 possible flips (None, horizontal or vertical). Then, the plankton transmission functions are individually modified so that $t_{plankton}(x, y) = \exp[-C \times a(x, y)]$ where C is a random constant and $C \in [0.5, 1]$. Next, the 13 transmission functions are randomly placed without overlapping on a 512×512 empty image to generate $t_0(x, y)$. Finally, the hologram $H_{z2}(X, Y)$ is simulated with Eq. 3 and Eq. 4. Both the holograms and the $t_0(x, y)$ are normalized between 0 and 1 and saved. 8,000 and 2,000 holograms were simulated for training and testing, respectively. Figure 3 presents an example of a simulated and labeled $t_0(x, y)$ and its corresponding hologram. During training, the object detection model learns to locate all the plankton on the raw holograms. The model should be able to predict the bounding boxes of the objects (x,y,w,h) and the class.

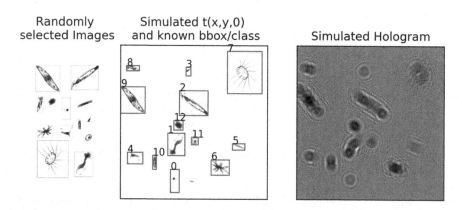

Randomly selected Images   Simulated t(x,y,0) and known bbox/class   Simulated Hologram

Fig. 3. Simulation example. 13 plankton images are used to generate a labeled hologram for a object detection task.

Tracking Dataset. To evaluate the tracking performances, we have simulated 100 videos that consist of 50 frames in which several plankton are moving in a 2D channel. For each simulated video, 10 plankton images were randomly selected from the ROIs used to test the detection model. For each selected plankton, we have simulated the transmission function $t_{plankton}(x, y) = \exp[-C \times a(x, y)], C \in [0.5, 1]$ which remained constant throughout the video. The plankton was then randomly placed on a 512×512 all-ones image with a non-overlapping constraint, so that the plankton does not initially occluded a previously placed plankton. Its velocity was then initialized with the calculation of the Poiseuille equation between two planes. Note that, for each video that lasts 50 frames, the 10 plankton are appearing or disappearing at different frame index according to their respective speed and frame of appearance (see Fig. 4).

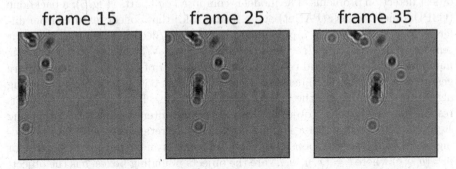

frame 15   frame 25   frame 35

Fig. 4. Tracking dataset example. 10 plankton are moving in a 2D channel.

For the simulations of the two datasets, we have considered $\lambda = 530$ nm (green), $z_2 = 0.8$ mm and a pixel size, which limits the final resolution, of 1.12 μm. The incident plane wave, described by a distribution $U(z) = \exp[i(k_x x + k_y y + k_z z)]$ where (k_x, k_y, k_z) are the wave vector components, was simulated with $U_0^-(x, y) = 1$ by choosing the position of the object at $z = 0$ and by selecting the optical axis along the propagation of the wave ($k_x = k_y = 0$) [6]. Note that since the plankton are placed close to the camera plane ($z_2 < 1$ mm), the simulated holograms are captured with a unit magnification [11]. As a result, the models trained on 512×512 images should be able to detect plankton over a field of view equal to 0.33 mm^2. The source code is available at https://github.com/romanescherrer/HoloTrack.

2.3 Object Detection Models and Tracking

Two tasks are to be considered in this paper. The first is the detection of objects on raw holograms which is performed frame by frame on a video. By detection, we mean the localization of all objects i.e. the determination of bounding boxes of coordinates (x, y, w, h) and the classification of objects (one among the considered 13 classes). The second task is the tracking of the objects throughout the video. This task, which aims at associating/linking detections across frames, allows, among other things, to determine the objects trajectories in order to precisely count the plankton that appear and disappear in the video without generating any duplicate.

Detection. To perform object detection task on raw holograms, we chose two YOLOv5 [3,15] models that were pre-trained on the COCO dataset, namely

YOLOv5s[2] (the smallest) and YOLOv5x (the largest) with $7.3M$ and $87.7M$ parameters, respectively. YOLO is a one-stage detector that integrates the detection of objects and their respective classification into a single process and has achieved state-of-the-art performances in term of speed and accuracy in many object detection problems. The model is composed of 3 parts (Fig. 5): a backbone (CSPDarknet), a neck (PANet) and a head (Yolo) that collect features from different stages of a $N \times N$ input images and encode/decode them into 3 output tensors of size $S \times S \times (B * (5 + n_c))$ where $S = (N/32, N/16, N/8)$, B is the number of anchors per grid cell and n_c is the number of classes. The anchors are generic bounding boxes dimensions (w,h) that are determined using a clustering algorithm (k-means) on the training dataset. Each cell in an output tensor is responsible for detecting objects within itself and after various post-processing steps (non-max suppression, among other, to only retain the candidate bounding boxes with higher response [3]), YOLO produces an output prediction vector $p = (b, o, c)$ where $b = (x, y, w, h)$ are the objects bounding boxes, o is the objectness i.e. a confidence score that the bounding boxes captures real objects and c is the class of the objects.

The models were trained on 8,000 holograms during 400 epochs with a batch size of 8 and tested on 2,000 holograms. The SGD optimizer was used with an initial learning rate equal to 0.01. To further evaluate the object detection performances on raw holograms, two models were also trained on the transmission functions $t_0(x, u)$ with the same hyperparameters. Note that $t_0(x, y)$ is the perfect image (artifact-free) that the holographic reconstruction steps seeks to obtain. Comparing the detection results on the holograms with those obtained on transmission function allow to determine whether the holographic reconstruction steps, which are iterative and time consuming, are avoidable to accurately classify and locate the objects with precision. The experiments were carried out on a 2.9 GHz Intel Core i7 PC with 64 GB of RAM and a Nvidia GTX 2060 GPU. The training took 8 h for the small model and 2 days for the larger one.

Tracking. Yolo is a real-time object detector [3] and thus can predict the bounding boxes and the classes of the objects at every frame of the video. In order to associate/link the detections across frames, we used the SORT algorithm proposed by [2]. The method works as follows (Fig. 6): During the algorithm initialization at the first frame noted k, each bounding box d_k detected by YOLO is associated with an unique tracker which is composed of a kalman filter. We denote $t_k\#n$ the bounding boxes of the trackers at the frame k where n is an unique identifier. For the next frame $k + 1$, the new bounding boxes d_{k+1} detected by YOLO must be associated to the existing trackers or new trackers must be created if the objects were not detected at the previous frame. For this, the kalman filters of the trackers predict the state of the bounding boxes at frame $k + 1$ by knowing the state of the bounding boxes at frame k. Then, the association of $d_{k+1}\#m, m \in [1, 2, ..., M]$ with the bounding boxes of

---

[2] https://github.com/ultralytics/yolov5.

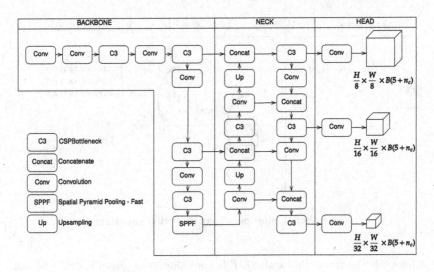

Fig. 5. YOLOv5 architecture.

the trackers $t_{k+1}\#n, n \in [1, 2, ...N]$ is performed by computing a cost matrix $C = \left(- IoU(d_{k+1}\#m, t_{k+1}\#n) \right) \in \mathbb{R}^{M \times N}$ where IoU is the Intersection over Union expressed by :

$$IoU(d_{k+1}, t_{k+1}) = \frac{d_{k+1} \cap t_{k+1}}{d_{k+1} \cup t_{k+1}} \tag{5}$$

The assignment is solved using the Hungarian algorithm and once a detection is associated to a target, the detected bounding box d_{k+1} is used to update the target state via the associated Kalman filter. The SORT algorithm is applied sequentially, frame by frame after the YOLO inference, on the whole video stream and the tracking can be done in real time because the state of the system at frame k is predicted by its previous state at frame $k - 1$.

2.4   Metrics

To evaluate YOLO, we report the object detection performances with the well-known average precision (AP) metrics [12]. We recall that the AP@.5 and AP@.75 are the average precision computed with an intersection over union threshold $t = 0.5$ and $t = 0.75$, respectively. The AP@[.5:.95] is reported by computing the mean AP@ with 10 different IoU thresholds [.5:.05:.95].

To evaluate the tracking performances, we report the CLEAR MOT metrics [1], with in particular:

– **MOTA**: The Multiple Object Tracking Accuracy metric that combines the false negative rate (FN), false positive rate (FP) and the mismatch rate ($IDSW$) into a single score :

$$MOTA = 1 - \frac{\sum_t (FN_t + FP_t + IDSW_t)}{\sum_t GT_t} \tag{6}$$

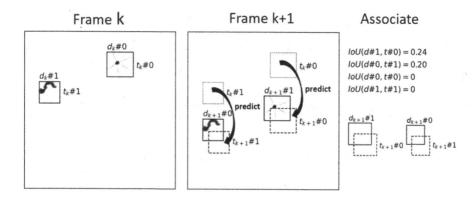

Fig. 6. SORT multiple object detection algorithm.

where t is the frame index and GT is the number of ground-truth objects.

- **MOTP**: The Multiple Object Tracking Precision that describes how precisely the objects are tracked by measuring and averaging the IoU between the objects and their corresponding hypothesis.

$$MOTP = \frac{\sum_{t,i} d_{t,i}}{\sum_t c_t} \tag{7}$$

where $d_{t,i}$ is the bounding boxes overlap between the target i and its assigned ground-truth objet and c_t is the number of matches.

- **MTR**: The Mostly Tracked Rate which is the percentage of ground-truth tracks that have the same label for at least 80% of their life span.
- **MLR**: The Mostly Lost Rate which is the percentage of ground-truth tracks that are tracked for less that 20% of their life span.

3   Results

3.1   Detection Performances

In this section, we report the object detection performances on 2000 test holograms and the mean inference time that includes FP16 inference, postprocessing and non-max suppression on a GTX 2060 GPU. Table 1 summarizes the performances of the object detection tasks performed on the raw holograms and on the transmission functions $t_0(x, y)$.

For the models trained on the holograms, the AP@.5 are 0.976 and 0.981 for YOLOv5s and YOLOv5x, respectively. For the models trained on the transmission functions, the AP@.5 are slightly better with 0.985 and 0.993 for YOLOv5s and YOLOv5x, respectively. The AP@[.5:.95] are significantly higher on $t_0(x, y)$ than on holograms (eg. 0.980 vs. 0.855 for YOLOv5x) but the AP@.75 are still high on holograms (0.928 and 0.955 for YOLOv5s and YOLOv5x, respectively). Those results suggest that the detectors trained on the holograms are efficient

Table 1. Detection performances.

Model	Inputs	AP@.5:.95	AP@.5	AP@.75	Speed
YOLOv5s	Holograms	0.820	0.976	0.928	4 ms
	$t_0(x, y)$	0.967	0.985	0.985	
YOLOv5x	Holograms	0.855	0.981	0.955	14 ms
	$t_0(x, y)$	0.980	0.993	0.989	

for a IoU threshold $\leq .75$ but that their performances start to decline at a higher threshold. Figure 7 shows the confusion matrix of YOLOv5x at IoU@.5 on the test holograms and an example of its predictions. One can notice that the diffraction pattern of an object spreads beyond its bounding box. In fact, the further away the object is from the camera, the more this effect will be visible on the hologram. Because of this and the lack of sharp edge, a detector trained on holograms was expected to have difficulty in determining the object boundaries with a high IoU.

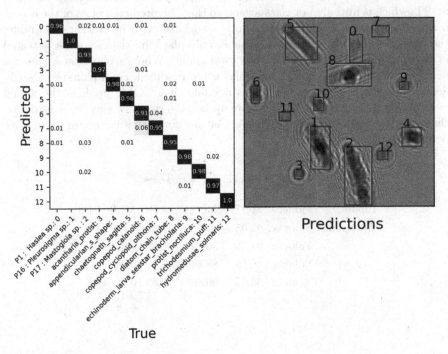

Fig. 7. Confusion matrix at IoU.5 and model predictions (blue: ground-truth, red: predicted). (Color figure online)

3.2   Tracking Performances

In this section we report the tracking performances performed by YOLO+SORT. We also computed the mean computation time of the whole pipeline when the real time detection (frame by frame) is performed by the smallest (v5s) and largest (v5x) versions of YOLOv5. The results of our evaluation are shown on Table 2. The pipeline YOLOv5s+SORT can be used up to 44 FPS while the extra large version can be used up to 23 FPS. This difference is explained by the fact that the model has to make its inference at each frame and that for very large models it is often more optimized in term of speed to generate predictions on a batch of observations. The results suggest that the performances difference between the small and large version of YOLO is negligible when the input images are $t_0(x, y)$. When the input images are holograms, the use of a larger model improves the performances but the number of lost tracks remains higher than that of the models trained on transmission functions. However, the tracking performance on holograms remains high with for example a MOTA of 94.34% and 92.03% for YOLOv5x and YOLOv5s, respectively. An exemplary output of our pipeline is shown in Fig. 8. At each frame of the video, the total number of plankton per species can be updated. Note that we have slightly modified SORT, which is initially not class-aware, so that the predicted class of the object is saved as soon as a YOLO detection is associated with its tracker. To update the plankton count by class at a frame k, only plankton that were not detected in the past frames are added to the total count. When a plankton leaves the field-of-view of the video, the total count is not modified. For a plankton already detected in the previous frames, it is possible that YOLO predicts the wrong class during its trajectory. We therefore update the count by class by considering that the detected object has the class that obtained the maximum occurrence between frames 0 to $k - 1$.

Table 2. Tracking performances.

Inputs	Model	MOTA	MOTP	MTR	MLR	FPS
Holograms	Yolov5s	92.03	84.76	92.54	1.94	44
	Yolov5x	94.35	86.33	95.30	1.43	23
$t_0(x, y)$	Yolov5s	96.16	88.89	96.32	0.72	–
	Yolov5x	96.05	90.66	96.63	0.92	–

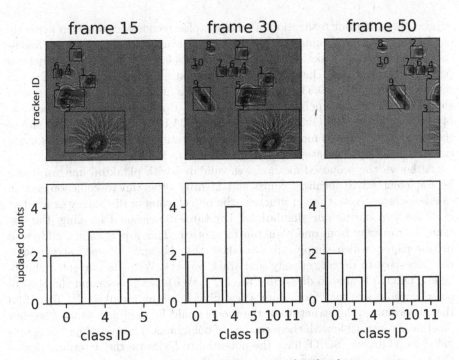

Fig. 8. Planckton tracked on a simulated video.

4   Conclusion and Perspectives

In this paper, we propose a pipeline that allows to detect, classify and count objects on raw holograms without going through the conventional holographic reconstruction/phase recovery steps. Our pipeline is composed of a real-time object detection model that performs the localization and classification of all objects present on the holograms and the SORT algorithm that links the detections through the video frames. We evaluated the object detection and tracking performances on simulated datasets that were generated with cropped plankton images obtained with a bright-field microscope and a shadow imager (ISIIS). Thirteen different species were considered for the simulations.

Two versions of YOLOv5 are trained to evaluate their detection performances on raw holograms. The results are compared with the detection performance obtained on transmission functions, which are the perfect images that the holographic reconstruction routine seeks to obtain. Note that in practice, obtaining a reconstructed holographic image of the same quality as our $t_0(x, y)$ in this paper is very complicated due to various noises and interferences on the hologram that can affect the conventional algorithms (focus/phase recovery) robustness. If anything, the presented comparison favors the holographic reconstruction/detection pipeline over the detection on raw hologram. However, although the results demonstrate that detection performances are slightly better on $t_0(x, y)$ than on holograms, the difference in AP@.5 is only 1.2%. These results suggest that the prior realization of a holographic reconstruction, even perfectly conducted, does not significantly increase the performance of the object classification and detection tasks. With a AP@.5 score of 0.981, a YOLOv5x model can perform detection and classification of all plankton groups within a 512×512 raw hologram (FOV ~ 0.33 mm^2) in a single pass in 14 ms. The tracking results show that the whole pipeline YOLOv5s+SORT can be performed in real-time (44 FPS) whereas YOLOv5x+SORT is slower (23 FPS) due to the large size of the model that required more floating-point operations suggesting that its usage could be more appropriate with batch (offline) tracking approaches.

Although the proposed method was validated with plankton images, it can be implemented to localize, count and identify other microscopic objects in raw holograms. Note that in practice, the object/camera distance was fixed at $z_2 = 0.8$ mm during our simulations. For three-dimensional imaging, the distance z_2 can vary from one plankton to another. This aspect is not addressed in this paper, which simply aims to show that holographic reconstruction is not necessary to detect, classify and track objects. With its current architecture, YOLOv5 is able to determine the (x, y, w, h) coordinates and the class of objects whose size may vary from a few pixels to a hundred pixels. To obtain the z-coordinate, the structure of the model could be modified. Otherwise, our pipeline is compatible with the recording of holograms. The bounding boxes provided by YOLOv5+SORT have the potential to facilitate the determination of the z-coordinate by any autofocusing algorithm.

While the results on simulated holograms are promising, it is often complicated and time consuming to put together a large dataset of real labeled holograms to train a detector. When a small labeled dataset is available, it might be beneficial to pre-train a detector with a large amount of simulated holograms and then use a transfer learning method to fine tune the model on the small dataset. Another approach would be to rely on an intensive data augmentation. Some works in the literature use de-focused back-propagated holograms as inputs of a deep learning model rather than raw holograms. By back-propagated the holograms on several planes near the correct global focus, the dataset could be significantly enlarged.

References

1. Bernardin, K., Stiefelhagen, R.: Evaluating multiple object tracking performance: the CLEAR MOT metrics. Eurasip J. Image Video Process. **2008** (2008). https://doi.org/10.1155/2008/246309
2. Bewley, A., Ge, Z., Ott, L., Ramos, F., Upcroft, B.: Simple online and realtime tracking. In: Proceedings - International Conference on Image Processing, ICIP 2016-August, pp. 3464–3468 (2016). https://doi.org/10.1109/ICIP.2016.7533003
3. Bochkovskiy, A., Wang, C.Y., Liao, H.Y.M.: Yolov4: optimal speed and accuracy of object detection. ArXiv:abs/2004.10934 (2020)
4. Lam, H.S., Tsang, P.W.: Invariant classification of holograms of deformable objects based on deep learning. In: IEEE International Symposium on Industrial Electronics 2019-June, pp. 2392–2396 (2019). https://doi.org/10.1109/ISIE.2019.8781149

5. Latychevskaia, T.: Iterative phase retrieval in coherent diffractive imaging: practical issues. Appl. Opt. **57**(25), 7187 (2018). https://doi.org/10.1364/ao.57.007187

6. Latychevskaia, T., Fink, H.W.: Practical algorithms for simulation and reconstruction of digital in-line holograms. Appl. Opt. **54**(9), 2424 (2015)

7. Lee, S.J., Yoon, G.Y., Go, T.: Deep learning-based accurate and rapid tracking of 3D positional information of microparticles using digital holographic microscopy. Exper. Fluids **60**(11) (2019). https://doi.org/10.1007/s00348-019-2818-y

8. Liu, X., Liu, X., Zhang, H., Fan, Y., Meng, H.: Research progress of digital holography in deep-sea in situ detection. Seventh Symposium on Novel Photoelectronic Detection Technology and Applications, vol. 11763, pp. 1760–1766 (2021)

9. Memmolo, P., et al.: On the holographic 3d tracking of in vitro cells characterized by a highly-morphological change. Opt. Express **20**(27), 28485–28493 (2012). https://doi.org/10.1364/OE.20.028485

10. Memmolo, P., et al.: Recent advances in holographic 3D particle tracking. Adv. Opt. Photon. **7**(4), 713 (2015). https://doi.org/10.1364/aop.7.000713

11. Mudanyali, O., Tseng, D., Oh, C.: Compact, light-weight and cost-effective microscope based on lensless incoherent holography for telemedicine applications. Lab. Chip **10**(11), 1417–1428 (2010). https://doi.org/10.1039/c000453g.Compact

12. Padilla, R., Passos, W.L., Dias, T.L., Netto, S.L., Da Silva, E.A.: A comparative analysis of object detection metrics with a companion open-source toolkit. Electronics (Switzerland) **10**(3), 1–28 (2021)

13. Persson, J., Mlder, A., Sven-Göran Pettersson, P., Alm, K.: Cell motility studies using digital holographic microscopy. Microsc. Sci Technol. Appl. Edu 4 (2010)

14. Picart, P., Montresor, S.: Digital Holography. Elsevier Inc. (2019). https://doi.org/10.1016/B978-0-12-815467-0.00005-0

15. Redmon, J., Divvala, S., Girshick, R., Farhadi, A.: You only look once: unified, real-time object detection. In: Proceedings of the IEEE Computer Society Conference on Computer Vision and Pattern Recognition 2016-Decem, pp. 779–788 (2016)

16. Rivenson, Y., Zhang, Y., Günaydin, H., Teng, D., Ozcan, A.: Phase recovery and holographic image reconstruction using deep learning in neural networks. Light Sci. Appl. **7**(2), 17141 (2018)

17. Scholz, G., et al.: Continuous live-cell culture imaging and single-cell tracking by computational lensfree LED microscopy. Sensors (Switzerland) **19**(5), 1–13 (2019). https://doi.org/10.3390/s19051234

18. Shimobaba, T., Kakue, T., Ito, T.: Convolutional Neural Network-Based Regression for Depth Prediction in Digital Holography. In: IEEE International Symposium on Industrial Electronics 2018-June, pp. 1323–1326 (2018)

19. Trujillo, C., Garcia-Sucerquia, J.: Automatic detection and counting of phase objects in raw holograms of digital holographic microscopy via deep learning. Opt. Lasers Eng. **120**, 13–20 (2019)

20. Wu, Y., et al.: Label-free bioaerosol sensing using mobile microscopy and deep learning. ACS Photon. **5**(11), 4617–4627 (2018). https://doi.org/10.1021/acsphotonics.8b01109

21. Wu, Y., Rivenson, Y., Zhang, Y., Günaydin, H., Lin, X., Ozcan, A.: Extended depth - of - field in holographic image reconstruction using deep learning based auto - focusing and phase - recovery. Optica **5**, 704–710 (2018)

22. Zhang, Y., Lu, Y., Wang, H., Chen, P., Liang, R.: Automatic classification of marine plankton with digital holography using convolutional neural network. Opt. Laser Technol. **139**(January), 106979 (2021)

The First *in-silico* Model of Leg Movement Activity During Sleep

Matteo Italia[1]([✉]), Andrea Danani[2], Fabio Dercole[1], Raffaele Ferri[3], and Mauro Manconi[4,5,6]

[1] Department of Electronics, Information, and Bioengineering, Politecnico di Milano, Milan, Italy
{matteo.italia,fabio.dercole}@polimi.it
[2] Dalle Molle Institute for Artificial Intelligence, University of Southern Switzerland, University of Applied Sciences and Arts of Southern Switzerland, Lugano Universitary Center, Lugano, Switzerland
andrea.danani@idsia.ch
[3] Oasi Research Institute – IRCCS, Troina, Italy
rferri@oasi.en.it
[4] Sleep Medicine Unit, Neurocenter of Southern Switzerland, Ospedale Civico, Lugano, Switzerland
[5] Faculty of Biomedical Sciences, Università della Svizzera Italiana, Lugano, Switzerland
mauro.manconi@eoc.ch
[6] Department of Neurology, University Hospital, Inselspital, Bern, Switzerland

Abstract. We developed the first model simulator of leg movements activity during sleep. We designed and calibrated a phenomenological model on control subjects not showing significant periodic leg movements (PLM). To test a single generator hypothesis behind PLM— a single pacemaker possibly resulting from two (or more) interacting spinal/supraspinal generators—we added a periodic excitatory input to the control model. We describe the onset of a movement in one leg as the firing of a neuron integrating physiological excitatory and inhibitory inputs from the central nervous system, while the duration of the movement was drawn in accordance with statistical evidence. The period and the intensity of the periodic input were calibrated on a dataset of subjects showing PLM (mainly restless legs syndrome patients). Despite its many simplifying assumptions—the strongest being the stationarity of the neural processes during night sleep—the model simulations are in remarkable agreement with the polysomnographically recorded data.

Keywords: Leg movement activity · Periodic leg movements · Restless legs syndrome

1 Scientific Background

Leg movement activity (LMA) during sleep refers to all tibialis anterior muscle activity events of one leg compliant with onset, offset, and amplitude criteria set

© The Author(s), under exclusive license to Springer Nature Switzerland AG 2022
D. Chicco et al. (Eds.): CIBB 2021, LNBI 13483, pp. 40–52, 2022.
https://doi.org/10.1007/978-3-031-20837-9_4

by the World Association of Sleep Medicine (WASM) [1]. LMA is detected by recording both tibialis anterior muscles by means of surface electromyography in the context of polysomnography. Periodic leg movements (PLM) are particular involuntary LMA, typically occurring during sleep (PLMS). PLM are a frequent phenomenon present in the majority of patients with restless legs syndrome (RLS), in a significant percentage of patients with other sleep disorders, and even in healthy subjects especially elderly [2,3].

Based on WASM criteria, PLM consist of series of at least four monolateral or bilateral candidate leg movements (CLM), each of them longer than 0.5 and shorter than 10 s (or 15 s for bilateral movements) and separated by 10–90 s. When two left and right movements overlap or are separated by less than 0.5 s they are considered as one bilateral leg movement; otherwise as two distinct monolateral movements [1]. The severity of PLM is quantified by the PLM index, indicating the PLM number per hour, and considered abnormal, by consensus, when it exceeds the value of 15 during sleep [4].

PLM might affect sleep quality for their association with cortical arousals and, in the long term, cardiovascular system, because of the repetitive induced increase of heart rate and blood pressure [5]. However, the mechanism and the neuroanatomic pathways behind PLM are largely unknown, as well as the origin of their periodicity. PLMS also occur in patients with complete transverse lesions indicating that the spinal cord contains the fundamental network to generate them [6,7]. In particular, since PLM might occur in one or both legs, one unsolved question is whether the network is generating one or two excitatory rythms (pacemakers).

The model is calibrated on data from both control and PLM subjects (subjects with significant/abnormal PLM index, in particular RLS patients). It allows to generate populations of virtual subjects, both control and PLM, and to simulate *in-silico* LMA. This goes beyond a speculative exercise: besides understanding the PLM cause, it has potential implications for the clinical practice. Indeed, this work is the first step in developing an *in-silico* laboratory that can bring tremendous benefits to doctors and patients, e.g., including pharmacological effects to investigate the fundamental decision to treat or not PLM [8], and, in case, how to optimally treat PLM. Moreover, a mathematical *in-silico* LMA model has indisputable advantages with respect to animal models, from both the ethical and economic viewpoints [9].

This work is a preliminary and modeling oriented version of the article recently appeared in the *Journal of Sleep Research* [10].

2 Materials and Methods

2.1 The LMA Model

We drastically simplify the underlying physiology (Figure 1 shows a schematic representation and an example of *in-silico* LMA generation) and assume that each leg is controlled by a single motor neuron (circular nodes in the figure),

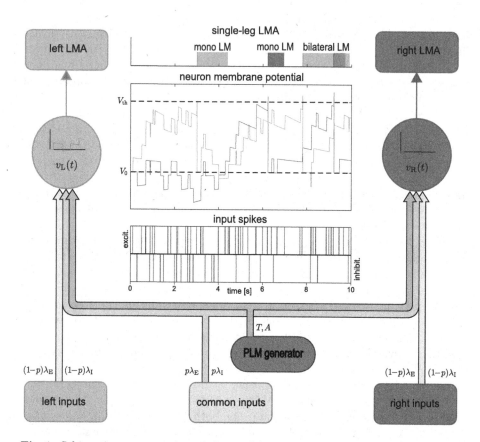

Fig. 1. Schematic representation of the model together with an example of *in-silico* LMA generation. The left and right motor neurons (red and blue circular nodes) implement the Stein integrate-and-fire model, with membrane potentials $v_L(t)$ and $v_R(t)$ at time t and the same rest potential V_0 and fire threshold V_{th} (see Eq. (1)). Leg specific and common physiological inputs from the central nervous systems (red, blue, and magenta rectangular source nodes) are modeled by spikes of synaptic current with Poissonian arrival times causing steps of equal amplitude a in the membrane potentials (positive/negative steps for excitatory/inhibitory spikes; arrival rates are indicated next to the input arrow). The assumed single PLM generator (brown) produces a periodic train of excitatory spikes with period T and potential step amplitude A. Example of simulated LMA: given the input spikes (stochastically drawn in accordance with their arrival rates), the left/right neuron membrane potential evolves according to Eq. (1); when the neuron fires, an LM starts on its controlled leg, with duration drawn from a data-fitted distribution (see in model calibration the Sect. 2.2.2). Single-leg LMA are combined in monolateral and bilateral LM according to the standard rule [1]: two monolateral and one bilateral LM (composed by one left LM and one right LM) are indicated in the top panel. (Color figure online)

representative of the central nervous system complex pathways ultimately determining contractions of the leg muscles. The two neurons are modeled with the
well-known Stein integrate-and-fire (IF) model [11] (see [12] for a review).

The IF mechanism is quite simple. The neuron state is characterized by its
membrane potential (middle in Fig. 1), which evolves according to the neuron
synaptic inputs. Excitatory (E)/inhibitory (I) inputs (bottom panel in Fig. 1)
increase/decrease the membrane potential (middle panel in Fig. 1), while a time
constant τ rules the potential discharge toward a resting value V_0. The neuron
is leaky since the summed contributions to the membrane potential decay with
τ. When the potential reaches a fire threshold V_{th}, an output spike is generated,
which causes the potential to reset at a basal value that we take, for simplicity,
to be the resting value V_0. The output spike represents, in our model, the leg
movement onset, while the duration of the movement is drawn from a distribution fitted on clinical data (see next section and the top panel in Fig. 1). We
neglect the neuron refractory period, i.e., after firing the neuron immediately
restarts integrating inputs (firing in the course of a LM is a rare event in our
model, $\leq 3\%$, and are disregarded with no significant effect on our results).

Each of the two neurons receive E/I inputs from the physiological activity of the
central nervous systems, a proportion p of which equally affect both legs (colored
in magenta in Fig. 1), while the remaining fraction $(1-p)$ is leg-specific (see the
rectangles and their firing rates in Fig. 1, red color for the left leg and blue for the
right one). For subjects showing significant PLM (subject typically characterized
by a bimodal distribution of the intermovement interval (IMI) [3,13] and by a large
PLM index), we add a periodic input common to both legs that implements our
hypothesis of a single phenomenological PLM pacemaker (brown PLM generator
in Fig. 1). The period and intensity of the periodic input are patient-specific.

We model the physiological inputs as series of synaptic current spikes, with
Poissonian arrival times, each causing the membrane potential to instantaneously
step by a small fraction a of the rest-to-fire interval $V_{th} - V_0$. We denote by λ_E
and λ_I the arrival rates of E and I inputs, divided into common (rate $p\,\lambda_X$)
and leg-specific (same rate $(1-p)\,\lambda_X$ for both legs, X $=$ E, I) (lower part of
Fig. 1). The three Poisson arrival processes (common and left-/right-specific)
are independent. The PLM pacemaker is a series of T-periodic synaptic spikes,
each spike causing an upward potential step A.

Denoting by $v_l(t)$ and $v_r(t)$ the left and right neurons membrane potential at
time t (red and blue curve in Fig. 1 middle panel), their time evolution is ruled
by the following ordinary differential equation:

$$\frac{d}{dt}v_x(t) = -\frac{v_x(t) - V_0}{\tau} + a\,S_E(t) + a\,S_{E,x}(t) - a\,S_I(t) - a\,S_{I,x}(t) + A\,S_P(t), \quad (1)$$

x $=$ l, r, where $S_E(t)$, $S_{E,x}(t)$, $S_I(t)$, and $S_{I,x}(t)$ are the series of unitary spikes of
the physiological inputs (common to both legs and leg-specific), and $S_P(t)$ is the
periodic series of unitary spikes of the PLM pacemaker (a unitary spike causing a
1-Volt upward step in the membrane potential). Between two consecutive inputs,
the potential $v_x(t)$ exponentially decays toward V_0 (the evolution is only ruled
by the discharge term in Eq. (1)). At the arrival of the new input spike, the
potential is updated by adding the spike contribution.

2.2 Model Calibration

Data were obtained from subjects previously enrolled in different studies on PLM published by some of the authors of this work (M.M. and R.F.) [6,14,15]. The LMA duration is fitted on all control LMA recordings (32 subjects characterized by unimodal IMI distribution; max PLM index = 5.9; mean age 48.03 years (SD 20.75) and 56.25% women). Parameters λ_E, λ_I, p, τ characterize control subjects. Their calibration result is a statistical fitting that can be used to generate virtual control subjects populations. The distributions of the PLM generator parameters T and A are fitted using the PLM dataset (65 subjects, mainly RLS affected, characterized by bimodal IMI distribution; PLM index > 15 except for a few cases; mean age 58.52 years (SD 13.09) and 66.15% women). All parameters are supposed constant during the night (8 h). All calibration procedures are implemented in Matlab. Figure 2 shows the results for the control model, whereas the calibration result of the PLM generator parameters is shown in Fig. 3. More details on the calibration follow.

2.2.1 Scaling Parameters

From a phenomenological standpoint, the membrane potential rest and threshold values, V_0 and V_{th}, are scaling parameters. We set $V_0 = 0$ and $V_{th} = 1$ and express the intensities a and A of the physiological and pathological synaptic inputs as fractions of the rest-to-fire interval (V_0, V_{th}, a, and A can therefore be considered adimensional). In particular, the intensity a played a scaling role affecting the calibration of the arrival rates λ_E and λ_I. With no loss of generality, we fixed $a = 0.1$ (($V_{th} - V_0)/10$), i.e., one-tenth of the rest-to-fire interval.

2.2.2 LMA Duration

We collected the durations of all LMA recorded on each single leg in the control dataset. We fitted the obtained LMA samples with all parametric distributions provided in Matlab. None however showed statistical agreement with the LMA data (Kolmogorov-Smirnov test (K-S) p-value < 0.01). We then used a non-parametric technique [16] (kernel density estimation, mean 2.47, SD 2.43, Fig. 2A; K-S p-value $\approx$ 0.15) and used the obtained distribution to independently draw the durations of all virtual LMA at the firing of the corresponding leg neuron. Because no significant correlation is documented between LMA duration and subject's PLM index, we use the LMA duration distribution fitted on controls also to generate the LMA of virtual PLM subjects.

2.2.3 Arrival Rates of Physiological Input Spikes

We jointly calibrated the arrival rates of the physiological input spikes on single leg recordings of the control subjects. Not distinguishing common from leg-specific inputs, E and I spikes arrive at rate λ_E and λ_I on each single leg, independently on the value of the proportion parameter p. The membrane time constant

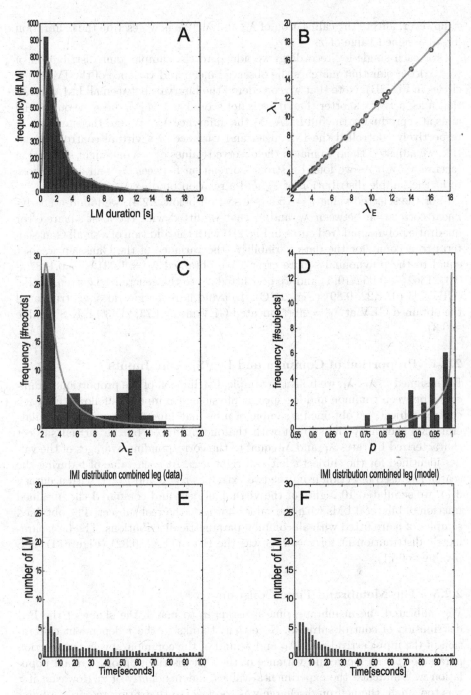

Fig. 2. Samples histograms (blue) and fitted distributions (red curve) of LMA duration (A), E inputs arrival rate λ_E (C), common/specific proportion parameter p (D). (λ_I, λ_E) correlation is shown in panel B. IMI distribution of the population of real (E) and virtual (F) control subjects. (Color figure online)

τ, however, affects the calibration of λ_E and λ_I. We now describe the calibration for an assigned value of τ.

For each single leg recording, we added to the sample joint distribution of (λ_E, λ_I) the pair that matches the observed mean and variance of the IMI (blue circles in Fig. 2B). Note that we considered the intervals between all LMA (recall that a leg activity shorter than 0.5 s is not scored as LM [1]), because our model aims at reproducing the full LMA. As the difference $\lambda_E - \lambda_I$ and the sum $\lambda_E + \lambda_I$ respectively controlled the IMI mean and variance of a virtual control subject [12], we adjusted them to match the observed values over a one-night simulation (accuracy 5%). As we found a strong correlation between λ_E and λ_I, we chose to fit the sample distribution of λ_E and a relation binding λ_I to λ_E.

For the selected value $\tau = 75$ s (see section below 2.2.5), we found a nearly-linear correlation between λ_E and λ_I that we fitted with the least-square-error quadratic polynomial (red curve in Fig. 2B) with the addition of a small Gaussian term to account for the data variability (the variance of the Gaussian was set equal to the polynomial square error). We obtained $\lambda_I = 1.031\lambda_E + 0.1187 + 0.002136\lambda_E^2 + 0.05\,N(0,1)$ and we best fitted λ_E to the generalized extreme value (GEV) (3.10, 1.22, 0.89) (Fig. 2B–C). To avoid unrealistic values, we truncated the obtained GEV at $\lambda_E = 20$ (truncated GEV mean 4.73, SD 3.14; K-S p-value ≈ 0.3).

2.2.4   Proportion of Common and Leg-Specific Inputs

For assigned τ, λ_E, λ_I, we built the sample distribution of the proportion parameter p between common and leg-specific physiological inputs as follows. For each control subject, we obtained a sample of p by matching the proportion of bilateral LMs shown by the subject with the one produced by the virtual subject characterized by rates λ_E and λ_I equal to the corresponding averages of the values identified for the subject's left and right legs. At each value of p during the search (we used bisection from the two extremes $p = 0$ and $p = 1$, with accuracy 3%), we simulated 10 nights of the virtual subject and compared the obtained fraction of bilateral LMs with the value shown by the real subject. The obtained samples of p are fitted with all common parametric distributions. The best parametric distribution fit for $\tau = 75$ s was the Beta (11.82, 0.82) (Figure 2D; K-S p-value ≈ 0.71).

2.2.5   The Membrane Time Constant

We calibrated the membrane time constant τ to match the shape of the IMI distribution of control subjects. Note that, thanks to the τ-dependent calibration of the input arrival rates λ_E and λ_I and of the common/specific proportion parameter p, the mean and variance of the IMI distribution of a virtual population well matched the experimental values, independently of τ. However, if τ was too small, the neuron discharge was too fast, so that firing required a burst of E inputs. This resulted in large input rates λ_E and λ_I and in a quite erratic dynamics of the neuron potentials, thus yielding a rather flat IMI distribution for the virtual population.

Since the characteristic IMI of control subjects is in the interval 0.5-10 s, we considered, as shape index, the ratio between the hourly number of IMI < 10 s, averaged over the population, and the average LMA index (the hourly number of LMA). The reason for considering the LMA index in the ratio, instead of the total hourly number of IMI, was that the input arrival rates were calibrated to match the subject LMA.

Summarizing, to evaluate a specific value of τ, we proceeded as follows: calibrate the distribution of λ_E and fit its correlation with λ_I; calibrate the distribution of p; generate a population of 100 virtual controls; simulate one night for each control and compute the IMI distribution shape index of the virtual population. As expected, the shape index increased with τ and got close to the experimental value (20%) at about $\tau = 75$. Figure 2E, F show the IMI distribution (restricted to the interval 0-100 s) of real and simulated control populations, respectively.

2.2.6 The Period of the PLM Generator

We relied on the fact that patients with significant PLM are characterized by a bimodal IMI distribution, where the first peak is typical of healthy subjects, while the second characterizes the PLM disorder [3,13]. We therefore built the sample distribution of the period T by taking the IMI of the second peak of each PLM subject's IMI distribution. We found the best fitting with the GEV (21.87, 3.49, 0.17) (left panel of Fig. 3; K-S p-value ≈ 0.55). Relaying on medical experience, we truncated the GEV below 17 and above 50 s (truncated GEV mean 24.42, SD 5.16; K-S p-value ≈ 0.59).

2.2.7 The Intensity of the PLM Generator

For each PLM subject, we drew 10 virtual control models, to each of which we added the PLM input with the subject-specific period T and amplitude A to be selected to match the subject's LM index. PLM subjects show more LMs than control ones, so that with $A=0$ the LM index, averaged on the 10 virtual subjects (each simulated for one night), falls below the value of the real subject. On the other extreme, if A is large, each PLM input spike triggers the firing of the neuron and the LM index of the virtual subjects exceeds the clinical value. We proceed via bisection to find the sample of A matching the subject's LM index (with accuracy of 3 movements/hr). We best fitted the LogNormal (0.78, 0.8) (K-S p-value ≈ 0.31; right panel of Fig. 3). To avoid unrealistic virtual PLM subjects, we truncated the distribution below 0.2 (truncated LogNormal mean 0.72, SD 0.60; K-S p-value ≈ 0.3).

3 Results and Discussion

Our model can be used to generate *in-silico* populations of both control and PLM subjects. We create populations of equal size to the datasets used for the model calibration (32 control and 65 PLM subjects) and compare the obtained sample distributions of typical clinical indicators against polysomnographically-recorded data. A limitation of this study is that we use for the comparison the same datasets used for calibration. On one hand, the datasets are too small to be split into calibration and validation. Moreover, the model parameters show a remarkable variability among the subjects, so that a correct validation would require more recordings of the same subjects. On the other hand, our model does not simulate the specific parameter values identified for single real subjects, but randomly draw virtual subjects from the statistics of the real control and PLM populations. Finally, the model is not aimed at forecasting the LMA of new subjects. Our primary aim in this work is to test the single-generator hypothesis behind PLM. For these reasons, it is acceptable to use the same dataset for calibration and assessment.

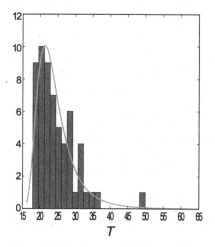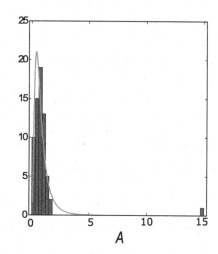

Fig. 3. Samples histograms (blue) and fitted distributions (red curve) of the PLM generator period T (panel left) and of the PLM generator intensity A (right panel). (Color figure online)

In calibration, we have identified the distributions of the model parameters that best fit the LMA of control and PLM subjects. In validation, we compare important features of virtual populations drawn from the identified statistics against the real populations.

Table 1 and Table 2 summarize the comparison, reporting mean and standard deviation of the principal LM features [14], durations, and composition of bilateral LM [15], together with their statistical agreement with real data, for control and PLM subjects respectively. We find a remarkable good agreement for LMA, LM, CLM, and the old PLM indeces [1,14] for both control and PLM populations. We also find accordance in the characteristics of monolateral and bilateral LM [15]. Noting that a virtual PLM subject is nothing but a virtual control subject with the only addition of the PLM periodic input, the latter being calibrated by fitting LMA indexes, rather than PLM indicators, we conclude that

Table 1. LM features in real (recording) and virtual (simulation) control subjects. Statistical agreement: T, Student's t-test; Tl, t-test on log-transformed data; U, Mann-Whitney U test.

	Control subjects (32)		
	Recording	Simulation	Statistics
LM features	**mean (SD)**	**mean (SD)**	**p-value**
LMA index	13.57 (±6.34)	13.74 (±6.45)	0.8804 (T)
LM index	13.44 (±6.29)	13.54 (±6.38)	0.9289 (T)
CLM index	12.88 (±5.58)	13.00 (±6.12)	0.9040 (T)
old PLM index	5.11 (±3.51)	4.34 (±1.48)	0.3611 (U)
PLM index	1.54 (±1.59)	0.65 (±0.73)	0.0193 (U)
Short-IMI index	2.71 (±2.15)	2.84 (±1.66)	0.4684 (U)
Mid-IMI index	4.89 (±2.71)	4.40 (±2.34)	0.6576 (U)
Long-IMI index	4.56 (±1.75)	5.02 (±1.70)	0.4051 (U)
Monolateral LMs	**median (IQR)**	**median (IQR)**	**p-value**
dunuration (s), min	0.53 (0.51–0.57)	0.52 (0.51–0.55)	0.9998 (T)
max	8.50 (6.42–9.72)	8.69 (7.53–9.29)	0.6255 (T)
mean	2.01 (1.76–2.28)	2.08 (1.93–2.21)	0.3178 (T)
median	1.59 (1.29–1.70)	1.54 (1.46–1.69)	0.4118 (T)
Bilateral LMs	**median (IQR)**	**median (IQR)**	**p-value**
# single LM, min	2.0 (2.00–2.00)	2.0 (2.00–2.00)	–
max	3.0 (2.00–3.00)	3.0 (2.00–3.00)	0.9158 (U)
mean	2.05 (2.00–2.11)	2.04 (2.00–2.10)	0.9561 (U)
median	2.0 (2.00–2.00)	2.0 (2.00–2.00)	–
duration (s), min	1.01 (0.82-1.13)	0.96 (0.81–1.22)	0.4573 (T)
max	9.19 (8.26–9.69)	9.67 (8.37–11.82)	0.0531 (T)
mean	3.91 (3.52–4.35)	4.09 (3.57–4.59)	0.2346 (T)
median	3.34 (2.98-3.91)	3.28 (2.85–3.74)	0.4109 (T)
Bilateral LMs (%)	31.45 (23.50–40.00)	30.94 (15.45)	0.8505 (T)

the statistical agreement gives support to the single-generator hypothesis behind the PLM phenomenon.

As expected, the agreement is strong in monolateral LM for control subjects, since the LM duration is fitted on all single-leg recordings of control subjects. Remarkably, the agreement remains very good also for the bilateral LM features validating the model for control subjects. Regarding PLM subjects, the obtained results support our model and thus the view of the single periodic generator. Indeed, not only the *in-silico* monolateral LM features statistically agree with the *in-vivo* ones, but also the bilateral ones, in particular the right increase of the proportion of bilateral LM in PLM subjects.

Table 2. LM features in real (recording) and virtual (simulation) PLM subjects. Statistical agreement: T, Student's t-test; Tl, t-test on log-transformed data; U, Mann-Whitney U test.

	PLM subjects (65)		
	Recording	Simulation	Statistics
LM features	**mean (SD)**	**mean (SD)**	**p-value**
LMA index	57.12 (±46.40)	61.36 (±45.48)	0.2436 (Tl)
LM index	56.90 (±46.22)	60.69 (±45.15)	0.2865 (Tl)
CLM index	55.31 (±42.67)	57.52 (±42.21)	0.5108 (Tl)
Old PLM index	44.22 (±32.02)	39.48 (±41.04)	0.0749 (U)
PLM index	29.17 (±19.78)	14.21 (±22.81)	≤ 0.0001 (U)
Short-IMI index	12.45 (±24.69)	17.15 (±11.67)	0.2573 (U)
Mid-IMI index	37.12 (±22.10)	35.08 (±31.61)	0.3899 (U)
Long-IMI index	4.05 (±1.76)	8.58 (±2.85)	≤ 0.0001 (U)
Monolateral LMs	**median (IQR)**	**median (IQR)**	**p-value**
dunuration (s), min	0.51 (0.51–0.53)	0.51 (0.51–0.52)	0.5943 (U)
max	9.30 (7.35–9.77)	9.42 (8.78–9.74)	0.5311 (U)
mean	2.15 (1.80–2.39)	2.06 (1.94–2.11)	0.3614 (U)
median	1.69 (1.40–2.14)	1.51 (1.42–1.57)	0.2269 (U)
Bilateral LMs	**median (IQR)**	**median (IQR)**	**p-value**
# single LM, min	2.0 (2.00–2.00)	2.0 (2.00–2.00)	–
max	3.0 (3.00–4.00)	4.0 (3.00–4.00)	0.1312 (U)
mean	2.03 (2.01–2.06)	2.05 (2.03–2.07)	0.0973 (U)
median	2.0 (2.00–2.00)	2.0 (2.00–2.00)	–
duration (s), min	0.92 (0.73–1.10)	0.83 (0.66–0.97)	0.1153 (U)
max	9.71 (9.47–9.91)	12.99 (11.56–13.87)	≤ 0.0001 (Tl)
mean	3.70 (3.34–4.31)	3.90 (3.67–4.09)	0.0502 (Tl)
median	3.32 (2.84–3.96)	3.18 (2.86–3.47)	0.1176 (Tl)
Bilateral LMs (%)	39.65 (28.11–53.90)	37.63 (29.96–50.52)	0.6965 (Tl)

The statistical agreement fails for the indicators requiring some temporal structure among LM. This is the case of the current PLM index [1] (Table 2). The PLM index considers only sequences of at least four consecutive LM separated by IMI in the interval 10–90 s and interrupted by IMI shorter than 10 s or longer than 90 s. The disagreement reason is rooted in the stationarity of the model parameters during the night, especially between the various sleep phases. Indeed, without requiring sequencing, we find a good agreement in the numbers of IMI in each of the three characteristic medical interval, i.e., 0–10 s (short-IMI, characteristic of healthy individuals), 10–90 s (mid-IMI, characteristic of PLM subjects), and 90 or more seconds (long-IMI). Only the number of long-IMI is larger in our simulations, but this is again an artefact of the model stationarity. Indeed, there is medical evidence that PLM decrease along the night, i.e., mid-IMI are concentrated at the beginning of the night, while few very long IMI characterizes phases with no PLM. However, calibrating a stationary PLM generator that matches, on average, the subject's number of LM (see calibration in Sect. 2.2) gives several long-IMI in lieu of less but longer ones.

4   Conclusion

We develop the first model to generate *in-silico* LMA, both for control and PLM subjects, adding only a PLM generator for PLM subjects. We calibrate the model parameters on recorded laboratory data and simulate control and PLM virtual populations. In spite of its simplicity, our phenomenological model shows a good statistical agreement between LMA features of *in-silico* and *in-vivo* populations. The agreement supports the validity of our model and also endorses the single generator hypothesis behind the PLM phenomenon. The main dissimilarities are caused by the model stationarity, opening up for future developments aimed at turning the model into a quantitative predicting tool to support medical intervention.

It is never easy to establish the merit of a first *in-silico* model of a physiopathological phenomenon, such as LMA, and to foresee possible future useful employment. However, the interested reader can find some ideas in section 4.1 Future research in [10], where we have tried to speculate on some possible future applications of our model, such as modeling the effects of drugs. E.g., by comparing the results between recordings in real PLM patients treated with dopamine-agonists (DA) and the corresponding virtual patients, the model might help in understanding if DA act on the neurological network of the periodic generator, or on another networks. Analogously, real and virtual datasets can be compared in other pharmacological circumstances, e.g., to confirm the so-far observed scarce effect on PLM of sedatives.

Considering also the patients' metadata, it will be possible to perform cluster analyses for the model parameters, favoring the important and still missing mission of PLM phenotyping.

Finally, the parameters A (PLM generator intensity) and T (PLM generator period) could be used as new indicators of the severity and temporality of PLM, respectively, to be used in parallel with the recently introduced parameters, like the periodicity index.

Acknowledgements. The authors thank the valuable suggestions received by the many reviewers. We also appreciate all the participants of the International Conference on Computational Intelligence Methods for Bioinformatics and Biostatistics, CIBB 2021, for their constructive feedback.

References

1. Ferri, R., et al.: World Association of Sleep Medicine (WASM) 2016 standards for recording and scoring leg movements in polysomnograms developed by a joint task force from the International and European Restless Legs Syndrome Study Groups (IRLSSG and EURLSSG). Sleep Med. **26**, 86–95 (2016)
2. Pennestri, M.H., Whittom, S., Adam, B., Petit, D., Carrier, J., Montplaisir, J.: PLMS and PLMW in healthy subjects as a function of age: prevalence and interval distribution. Sleep **29**(9), 1183–1187 (2006)
3. Ferri, R.: The time structure of leg movement activity during sleep: the theory behind the practice. Sleep Med. **13**(4), 433–441 (2012)
4. American Academy of Sleep Medicine: International classification of sleep disorders (2014)
5. Ferri, R., Koo, B., Picchietti, D.L., Fulda, S.: Periodic leg movements during sleep: phenotype, neurophysiology, and clinical significance. Sleep Med. **31**, 29–38 (2017)
6. Ferri, R., et al.: Neurophysiological correlates of sleep leg movements in acute spinal cord injury. Clin. Neurophysiol. **126**(2), 333–338 (2015)
7. Salminen, V., et al.: Disconnection between periodic leg movements and cortical arousals in spinal cord injury. J. Clin. Sleep Med. **9**(11), 1207–1209 (2013)
8. Figorilli, M., Puligheddu, M., Congiu, P., Ferri, R.: The clinical importance of periodic leg movements in sleep. Current Treat. Opt. Neurol. **19**(3), 10 (2017)
9. Allen, R.P., et al.: Animal models of RLS phenotypes. Sleep Med. **31**, 23–28 (2017)
10. Italia, M., Danani, A., Dercole, F., Ferri, R., Manconi, M.: A calibrated model with a single-generator simulating polysomnographically recorded periodic leg movements. J. Sleep Res. **31**(5), e13567 (2022)
11. Stein, R.B.: A theoretical analysis of neuronal variability. Biophys. J. **5**(2), 173–194 (1965)
12. Burkitt, A.N.: A review of the integrate-and-fire neuron model: I. Homogeneous synaptic input. Biol. Cybern. **95**, 1–19 (2006)
13. Ferri, R., Zucconi, M., Manconi, M., Plazzi, G., Bruni, O., Ferini-Strambi, L.: New approaches to the study of periodic leg movements during sleep in restless legs syndrome. Sleep **29**(6), 759–769 (2006)
14. Ferri, R., et al.: Putting the periodicity back into the periodic leg movement index: an alternative data-driven algorithm for the computation of this index during sleep and wakefulness. Sleep Med. **16**, 1229–1235 (2015)
15. Ferri, R., et al.: Bilateral leg movements during sleep: detailing their structure and features in normal controls and in patients with restless legs syndrome. Sleep Med. **32**, 10–15 (2017)
16. Gramacki, A.: Nonparametric Kernel Density Estimation and its Computational Aspects. Springer (2018)

Transfer Learning and Magnetic Resonance Imaging Techniques for the Deep Neural Network-Based Diagnosis of Early Cognitive Decline and Dementia

Nitsa J. Herzog[1]([envelope]) [iD] and George D. Magoulas[2] [iD]

[1] Department of Computer Science, Birkbeck College, University of London, London WC1E 7HZ, UK
nitsa@dcs.bbk.ac.uk
[2] Birkbeck Knowledge Lab, University of London, London WC1E 7HZ, UK
g.magoulas@bbk.ac.uk

Abstract. Combining neuroimaging technologies and deep networks has gained considerable attention over the last few years. Instead of training deep networks from scratch, transfer learning methods have allowed retraining deep networks, which were already trained on massive data repositories, using a smaller dataset from a new application domain, and have demonstrated high performance in several application areas. In the context of a diagnosis of neurodegenerative disorders, this approach can potentially lessen the dependence of the training process on large neuroimaging datasets, and reduce the length of the training, validation, and testing process on a new dataset. To this end, the paper investigates transfer learning of deep networks, which were trained on ImageNet data, for the diagnosis of dementia. The designed networks are modifications of the AlexNet and VGG16 Convolutional Neural Networks (CNNs) and are retrained to classify Mild Cognitive Impairment (MCI), Alzheimer's disease (AD) and normal patients using Diffusion Tensor Imaging (DTI) and Magnetic Resonance Imaging (MRI) data. An empirical evaluation using DTI and MRI data from the ADNI database supports the potential of transfer learning methods in the detection of early degenerative changes in the brain. Diagnosis of AD was achieved with an accuracy of 99.75% and a 0.995 Matthews correlation coefficient (MCC) score using transfer learning of VGG models retrained on DTI scans. Early cognitive decline was predicted with an accuracy of 93.88% and an MCC equal to 0.8602 by VGG models processing MRI data. The proposed models can be used as additional tools to support a quick and efficient diagnosis of MCI, AD and other neurodegenerative disorders.

Keywords: MRI · DTI · Transfer learning · Dementia · Deep learning

1 Introduction

Mild cognitive impairment (MCI) belongs to the group of neurocognitive disorders characterized by minor problems with cognitive function, including memory, language,

D. Chicco et al. (Eds.): CIBB 2021, LNBI 13483, pp. 53–66, 2022.
https://doi.org/10.1007/978-3-031-20837-9_5

visual and spatial perception. Detecting MCI early is important since approximately 15% of the 65-year-olds with MCI develop dementia within a year, and 30% of them develop it within 5 years. The most common course of dementia is Alzheimer's disease (AD). Neuroimaging technology is one of the key diagnostic approaches for the detection of early dementia. In this context, Magnetic Resonance Imaging (MRI) scans give detailed characteristics of the anatomical properties of the brain and cover around 50% of imaging data used for the diagnosis of brain diseases [1]. Also, Diffusion Tensor Imaging (DTI) provides the complex anatomy of the fiber tracts at the microstructural level and creates a brain-wide mapping of neuronal connections between the anatomical regions [2]. Both methods are widely used in the diagnosis of MCI and AD. Previous research has pointed out that in the early phases of the disease, white matter (WM) tract damage is happening earlier than gray matter (GM) destruction and the progression of WM atrophy exceeds the grey matter degeneration in patients with dementia [3, 4]. It has been highlighted that there is a significant correlation between WM changes and regional GM atrophy in patients with AD and this affects cognitive test performance [5]. At the same time, the correlation between GM atrophy and the damage of most WM tracts was not found in patients with the amnestic forms of MCI. In this vein, the study presented in this paper uses both imaging techniques for the early diagnosis of dementia.

In the last decade, a significant number of studies used machine learning methods for medical diagnosis [6, 7], with support vector machines (SVM), support vector regression (SVR), and random forest (RF) classifiers being among the most popular methods [6]. Advances in deep neural networks have opened a wide diagnostic opportunity in the classification and processing of medical imaging data offering additional benefits [7], and among them, Convolutional Neural Networks (CNNs) have demonstrated great potential in medical image analysis [8].

Transfer learning, which is at the core of this paper, became noticeable in medical diagnostics only in recent years. Its popularity is growing as it is a fast and highly effective approach [9]. Although deep neural networks can learn various combinations of features from coarse to fine, they typically require a lot of training data and specialized computing infrastructure to do so. Transfer learning strategies are based on thoroughly trained deep networks which are retrained using a smaller dataset from a new application domain. A popular strategy to implement transfer learning of pretrained networks is replacing the last three layers of the network's architecture. This allows adjusting the existing network to the requirements of a new classification domain and has demonstrated in several applications at least comparable performance with models trained from scratch.

The paper explores the classification potential of popular CNN architectures, such as the AlexNet and the VGG16 networks, which have been trained on ImageNet data (www.image-net.org). Transfer learning enables quick adaptation of these computational models to new classes of medical imaging data from MRI or DTI with minimal image preprocessing. The aim is to understand how transfer learning with deep networks can be used to inform the design of DTI or MRI-based diagnostic tools for binary or multiclass classification of early mild cognitive impairment, Alzheimer's disease and Normal (healthy) Controls (NC). This approach could offer new opportunities for quick and efficient diagnostics of different medical conditions including neurodegenerative disorders.

2 Deep Learning for Medical Diagnosis

Compared to the classical machine learning algorithms, deep neural networks can provide an end-to-end solution, automating the image preprocessing and feature engineering stages, by considering those as part of the training process, and are able to achieve a high prediction rate of brain pathology. For example, deep learning can be used as a single classifier or in ensemble architectures for the diagnosis of brain degenerative diseases [10]. Deep networks can handle 2D and 3D data in order to distinguish between healthy and dement subjects [11]. At the same time, the advantages of deep learning models can be used for limited datasets by applying a layer-wise transfer learning approach [12] and image augmentation techniques [13]. Deep transfer learning models propose an effective way of image segmentation and can automatically classify brain scans focusing only on small brain regions [14]. Solutions proposed so far were tested only one imaging technology with very few attempts to classify Alzheimer's Disease using joint sets of MRI plus DTI data [14]. The current research complements this effort by focusing on the classification comparison of two transfer learning models. That is performed on two imaging technologies which are tested separately to determine a suitable image-algorithm combination for distinguishing between dementia stages.

2.1 Convolutional Neural Network for Image Classification

Convolutional Neural Networks, which are the center of our transfer learning scheme, are a class of multilayer neural networks that adopt the Deep Learning paradigm [8]. Input and output layers are tensors and are connected via several hidden layers of weighted nodes. Hidden layers perform important functions, data transformations, calculations, and analyses. The weights are learned and adapted by optimization algorithms. All layers are chained together. The output layer collects the processed information and generates the output, which can represent a prediction or a categorization depending on the application context.

The CNN architecture was specially designed for imaging data [15]. In that case, a two-dimensional grid of pixels typically represents each image. Each pixel value and location might be associated with numerical values depending on the black-and-white, grayscale or color images. While processing imaging data, a neural network architecture must follow the relevant application requirements. The first of these requirements is translational invariance. It means that the network layers should respond similarly to the same area regardless of where it appears in the image. The second requirement is based on the principle of "locality" when the earliest layers concentrate mainly on local regions, simple features, and abstractions. The local area representations can aggregate knowledge about the whole image.

Convolution Functions

Image processing with multilayer neural networks typically requires transforming images into one-dimensional vectors. This kind of conversion impacts the relationship between the image pixels and makes a neural classifier less effective in image processing, requiring a high number of parameters and extensive training time. In contrast, CNN can receive a tensor at the input and can learn spatial relations between pixels of the image.

One can exploit the benefits of CNNs by designing significantly deeper neural architectures which, nevertheless, can learn fast complex relations from raw images. This allows the CNN to detect useful features automatically during training and develop an internal representation that classifies more complicated images than a normal multilayer neural network with sigmoid activations.

The main structural element of a CNN is the convolutional layer that operates using a convolutional function [16]. The convolution between two functions, measuring the overlap between f and g as a function of X, can be defined as:

$$(f * g)(X) = \int f(z)g(X - z)dz, \tag{1}$$

when one of the functions is flipped and shifted by the distance z.

For discrete objects such as, for instance, a set of infinite-dimensional vectors, the formula takes the following form:

$$(f * g)(i) = \sum_a f(a)g(i - a). \tag{2}$$

In the case of a 2D tensor, such as those used when for imaging, a corresponding sum with indices (a, b) will look as follows:

$$(f * g)(i, j) = \sum_a \sum_b f(a, b)g(i - a, j - b). \tag{3}$$

Pooling Layer

Pooling layers help to save a global image representation by keeping all the advantages of the convolutional layers and other intermediate layers [16]. At the same time, the pooling procedure makes the image size significantly smaller and might alleviate the overfitting problem of the entire neural network.

The pooling operators are deterministic. They usually compute average or maximum values and are called "averaging pooling" or "max pooling" respectively. These average or maximum values are calculated at each layer location depending on the pooling function employed. The pooling layer significantly reduces the network layers' size keeping the most significant spatial layer information, in an attempt to reduce overfitting.

ReLU Activation Layer

The activation layer uses differentiable operators to transform the weighted sum of the inputs received by a neuron to outputs. There are several popular activation functions for deep neural networks, such as Rectified linear unit (ReLU), sigmoid, hyperbolic tangent (tanh), and Softmax [17].

The ReLU helps the neuronal network to learn fast and produce good performance [18]. For a given element x, the function can be expressed as the maximum of that element x and 0:

$$ReLU(x) = \max(x, 0). \tag{4}$$

Dropout Layer

Dropout is an efficient way to prevent the neural network from overfitting by applying a regularization technique [19]. A Dropout layer randomly sets some inputs to zero at each update of the training circle, reducing the network's capacity. All other inputs are scaled up to 1 such that the sum of all inputs remains the same.

The dropout technique can be applied to most layered neural architectures, such as models with convolutional layers, long short-term memory layers, recurrent layers, and fully connected layers. Dropout can be applied to the input layer in some situations, but it is never used with the output layer. The most advanced dropout technique specifies the probability at which parameters perform the dropout procedure. A standard threshold value for the retaining output is a probability of 0.5 for each hidden node of the layer, which means that the network retains all values above this level. The role of the dropout probability p can be explained with the following formula where each intermediate activation of the hidden node H is replaced by the random variable H':

$$H' = \begin{cases} 0 & \text{with probability } p \\ \frac{H}{1-p} & \text{otherwise.} \end{cases} \tag{5}$$

After the dropout procedure, the network weights will become larger than before. Therefore, the weights are usually scaled between zero and one before saving the model.

In summary, the CNN layers have the following functions: convolutional layers are used for feature engineering, pooling layers reduce the dimensions of the feature maps, activation layers normalize the feature maps by removing the negative values, output layers produce the classification result, the dropout layer reduces the model overfitting, and the fully connected layers compute a score of each class collected from convolutional layers.

2.2 Pretrained Convolutional Neural Network

Several popular CNN architectures have been used as base models in research projects and incorporated into modern machine learning packages. Most of them were originally introduced in the context of the ImageNet competition, launched in 2010, and won the first prize. ImageNet is the main forum for demonstrating advances in new supervised learning models in the area of computer vision. The performance of the pretrained models varies depending on the architecture and the choices of hyperparameters. A common transfer learning strategy consists of using pre-trained layers to construct a different network that might have similarities to the first layers. Pretrained models are available in many interesting configurations that can be grouped according to their architectural similarities. One of the first successful models is the AlexNet network. Another group of models is the Visual Geometry Group (VGG) networks, which were created from repeating blocks of structural elements that were originally introduced in the VGG model - a model built to detect geometric shapes. Another group is based on the GoogLeNet and differs from the previously mentioned architectures in the use of "Inception" blocks, which consist of parallel convolutional layers with filters of different sizes and max-pooling layer whose outputs are concatenated. Other approaches include the Network in Networks (NiN) which is based on small patch-wised convolutions, the Residual

Networks (ResNet), which consist of different numbers of residual blocks and channels, and the Densely connected networks (DenseNet) that extend further the concept of residual blocks introduced in ResNet. An overview of the two models more relevant to this work is presented next.

AlexNet

AlexNet has been named after the first name of his developer, Alex Krizhevsky, who won the ImageNet competition in 2012 [20]. ImageNet competitors trained models on one million images of one thousand object categories. The lowest layers of the model are supposed to detect edges, texture, and colors, resembling the traditional image filters. The hidden layers learn a compact representation of the image whose property can be easily separated into the different data categories.

AlexNet requires an input image size of 224 × 224 pixels. The network was trained using an image augmentation approach, such as image clipping, flipping, and usage of color channels, which makes the model more robust, reducing overfitting. The AlexNet consists of eight layers, including five convolutional, two fully connected hidden, and one fully connected output layer. The network does not require manually designed features. All the feature detection and extraction procedures are done automatically. The first convolution filter (window) has a size of 11 × 11. It gives the possibility to capture rather big objects. The second and third convolutional layers have reduced filter sizes of 5 × 5 and 3 × 3, respectively.

Furthermore, the network has max-pooling layers inserted after the first, second, and fifth convolutional layers. All max-pooling layers have a window size of 3 × 3 and slide through the layers with a stride of 2. AlexNet uses ReLU activation functions. The network architecture is completed by two fully connected layers of 4096 output parameters (8192 parameters in total divided between dual GPUs). The model complexity is controlled by adding the dropout layer to the fully connected layer.

VGG

A repeated block structure characterizes the VGG network architecture [21]. It can be divided into two parts- integrated convolutional blocks and several fully connected layers. Each building block includes a sequence of convolutional layers with the kernel of size 3 × 3 and padding of 1 pixel, a max-pooling layer of size 2 × 2 and a stride of 2 pixels. The original VGG network (VGG-11) has five blocks of convolutions and 11 layers overall. The first two blocks of the network have one convolutional layer each, and the following three blocks include two convolutional layers each. The number of output channels of the first block is 64. This number doubles with each successive block and reaches 512 in the final one. Like the AlexNet, the VGG uses the ReLU activation function. Output parameters on the fully connected layer are equal to 4096, 4096 and 100 respectively. Compared to the AlexNet, the VGG-16 is computationally heavier. The VGG network has several modifications with exceeded number of convolutional layers, e.g., the VGG-16 and the VGG-19.

The two pretrained networks, AlexNet and VGG, are retrained and tested in the current research, as discussed in Sect. 4.

3 Imaging Data Repositories

Brain scans were obtained from the Alzheimer's Disease Neuroimaging Initiative (ADNI) database (adni.loni.usc.edu) - a well-known repository of neuroimaging data. The created datasets include T1-waited images of structural MRI and DTI data of fractional anisotropy of 150 subjects. EMCI and NC patients were between 55 and 65 years old, whilst AD patients were between 65 and 90 years old. Images were processed and classified in Matlab using commodity hardware (Windows10 Enterprise, Intel (R) Core (TM), i7–7700 CPU@ 3.60 GHz, 16 GB RAM).

4 Proposed Transfer Learning Pipeline

Initially, MRI and DTI datasets of 2D images from the ADNI3 database were created. Images were taken from the same type of 3T scanners, Siemens Medical Solutions (see details available on ADNI: http://adni.loni.usc.edu/methods/mri-tool/mri-acquisition). For the MCI and NC classes, patient data from the age group 55 to 65 years old were used to minimize the ageing effect on the imaging data. The MRI and DTI brain images were normalized using the histogram stretching technique and resized to 256 × 256 pixels with RGB color channels as typically done for deep learning image processing and classification. Then, the brain area in a single 2D image was segmented from the skull and other surrounding tissues using region growing and double thresholding methods (see Fig. 1). A set of 600 MRI and 600 DTI images (4 slices from each subject) were obtained from the 150 subjects. Images were balanced across classes and were used for binary and multiclass diagnosis.

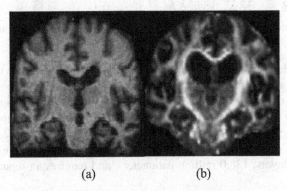

(a)   (b)

Fig. 1. (a) Segmented brain from MRI slice (b) Segmented brain from DTI slice.

The classification tasks were processed using transfer learning of two CNN architectures, the AlexNet and the VGG16, where the last three layers were replaced by a fully connected layer, a Softmax activation layer and an output layer, which was configured for binary or multiclass (three classes) classification depending on the type of the diagnostic task. When the cross-entropy loss function is used for training, the outputs of the Softmax layer can be interpreted as values of a probability distribution, which helps to produce the diagnostic outcome.

The weights of the pretrained AlexNet and VGG16 are used as parameters when adapting the pretrained models to the new task by retraining them using the MRI and DTI sets. Figure 2 illustrates how transfer learning of these models was used for the diagnosis of Mild Cognitive Impairment and Alzheimer's Disease; notice that the last three layers were replaced to adjust each model to the application domain.

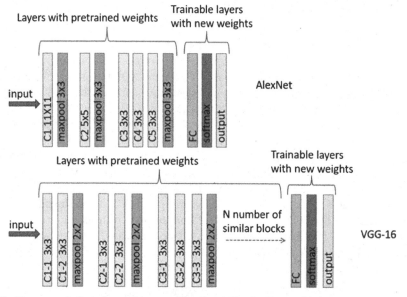

Fig. 2. Deep transfer learning architectures for diagnosis of early cognitive decline and dementia.

In general, AlexNet has been found to provide a short training time, while the VGG16 has been proved able to produce low error rates. AlexNet and VGG16 were originally configured and trained for 1000 classes using ImageNet data. AlexNet consists of 8 layers, has a size of 227MB and includes 61.0 million parameters. As mentioned in the previous section, this network requires an input image size of $227 \times 227 \times 3$ (227 wide, 227 high, 3 color channels). The size of VGG16 is much bigger, reaching 515MB. This network has 16 layers, 138.0 million parameters, and requires an input image size of $224 \times 224 \times 3$.

The following setup was used for retraining/finetuning both models on DTI and MRI data. All brain images were resized to fit the two pretrained networks' input sizes and were fed into the models: 80% of the images were used for training, 10% for validation, and an independent 10% of images were used for testing. Training parameters were set as follows: $N = 5$ for the number of training epochs for each dataset; mini-batch size = 128; validation data frequency = 50. The stochastic gradient descent with momentum (SGDM), with an initial learning rate = 0.0001, was used to train the models. All the results below are presented for the test MRI and DTI image data.

5 Experiments and Results

Experiments were conducted with the adapted configurations of AlexNet and VGG16, as described in Sect. 4, using DTI and MRI data. Four classification problems were tested: three binary classification tasks (EMCI vs. NC, AD vs. NC, and AD vs. EMCI) composed of 400 images each, and one multiclass task (AD vs EMCI vs NC) using 600 images with a balanced number of AD, EMCI, and NC subjects.

Table 1. Average classification performance (over 25 independent runs) on test DTI test data using transfer learning with VGG16

Model	Multiclass AD, EMCI, NC	Binary AD vs EMCI	Binary EMCI vs NC	Binary AD vs NC
Datasets	DTI	DTI	DTI	DTI
Acc	0.8438 ± 0.020	0.7400 ± 0.030	0.9100 ± 0.015	0.9975 ± 0.001
Precision	0.8600 ± 0.023	0.7450 ± 0.024	0.9200 ± 0.010	1.0000 ± 0.000
Recall	0.8329 ± 0.030	0.7376 ± 0.018	0.9020 ± 0.019	0.9950 ± 0.005
F-score	0.8462 ± 0.017	0.7413 ± 0.034	0.9109 ± 0.011	0.9975 ± 0.002
Specificity	0.7975 ± 0.018	0.7234 ± 0.031	0.9020 ± 0.012	0.9950 ± 0.003
MCC	0.6899 ± 0.021	0.4781 ± 0.019	0.8202 ± 0.019	0.9950 ± 0.005
AUROC	0.9766 ± 0.010	0.8581 ± 0.012	0.9700 ± 0.010	0.9998 ± 0.001

Table 2. Average classification performance (over 25 independent runs) on MRI test data using transfer learning with VGG16

Model	Multiclass AD, EMCI, NC	Binary AD vs EMCI	Binary EMCI vs NC	Binary AD vs NC
Datasets	MRI	MRI	MRI	MRI
Acc	0.8950 ± 0.034	0.7813 ± 0.015	0.9300 ± 0.016	0.9350 ± 0.020
Precision	0.8900 ± 0.028	0.7950 ± 0.025	0.9200 ± 0.020	0.9200 ± 0.028
Recall	0.8990 ± 0.025	0.7737 ± 0.027	0.9388 ± 0.020	0.9485 ± 0.019
F-score	0.8945 ± 0.017	0.7842 ± 0.019	0.9293 ± 0.025	0.9340 ± 0.020
Specificity	0.8641 ± 0.019	0.7626 ± 0.030	0.9388 ± 0.016	0.9604 ± 0.017
MCC	0.7922 ± 0.023	0.5630 ± 0.022	0.8602 ± 0.020	0.8710 ± 0.023
AUROC	0.9800 ± 0.011	0.8787 ± 0.012	0.9800 ± 0.010	0.9756 ± 0.013

Tables 1 and 2 summarize the models' classification performance in testing, after applying the transfer learning process described in Sect. 4 for the VGG network. Twenty-five independent runs were conducted by repeating the training process with a different random seed in each case. The metrics shown include accuracy rate (Acc), precision, recall, specificity, the area under the curve, which plots parametrically the true positive rate vs the false positive rate (AUC), and the F-score, which is commonly used for evaluating the performance of machine learning models. It is defined as the harmonic mean of the model's Precision and Recall (see Tables 1 and 2). Lastly, the Matthews correlation coefficient (MCC) is also reported as a metric of the quality of classifications which measures the correlation of the true classes with the predicted labels for binary classification tasks. MCC for multiclassifier was computed from averaging values of true positive (TP), true negative (TN), false positive (FP) and false negative (FN) results.

The highest performance is achieved for the binary datasets that include the AD and NC classes. Metrics for the diagnosis of EMCI vs. NC also reveal positive performance with MRI test images diagnosed more accurately than DTI test data. Distinguishing between AD and EMCI is a more challenging task, and the VGG models perform lower compared to the other binary tasks with a better performance on MRI images. Multiclassification accuracy varies from 84% for DTI data to 89.5% for MRI images. Figure 3 visualizes the classification quality as represented by the MCC for DTI and MRI data for multi- and binary classes.

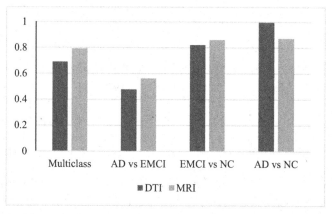

Fig. 3. VGG-based transfer learning MCC score in multiclass and binary classification of DTI and MRI data.

It is worth noticing that, as shown in Fig. 3, transfer learning with the VGG classifier detects early cognitive decline with an MCC of 0.82 using DTI images and 0.86 when MRI scans are used.

Tables 3 and 4 summarize classification performance using transfer learning with the AlexNet.

As with the VGG classifier, the classification quality between images of the AD and EMCI classes is lower compared to EMCI (EMCI vs NC) or AD (AD vs NC) class data. The classification performance for the detection of early brain changes (EMCI vs NC) can vary but without significant differences between DTI and MRI data. The diagnosis of Alzheimer's Disease using DTI data is significantly higher (99%) than using MRI data (86%) for transfer learning with the AlexNet. Comparative results in terms of MCC score in binary and multiclass classification of DTI and MRI data are shown in Fig. 4.

Table 3. Average classification performance (over 25 independent runs) on test DTI data using transfer learning with AlexNet

Model	Multiclass AD, EMCI, NC	Binary AD vs EMCI	Binary EMCI vs NC	Binary AD vs NC
Datasets	DTI	DTI	DTI	DTI
Acc	0.7088 ± 0.025	0.6900 ± 0.023	0.8500 ± 0.025	0.9900 ± 0.010
Precision	0.7000 ± 0.034	0.7200 ± 0.027	0.8700 ± 0.019	0.9900 ± 0.010
Recall	0.7125 ± 0.020	0.6792 ± 0.025	0.8365 ± 0.025	0.9900 ± 0.010
F-score	0.7062 ± 0.018	0.6990 ± 0.020	0.8529 ± 0.250	0.9900 ± 0.010
Specificity	0.6731 ± 0.015	0.6800 ± 0.023	0.8163 ± 0.028	0.9895 ± 0.013
MCC	0.4128 ± 0.012	0.3801 ± 0.018	0.7008 ± 0.030	0.9799 ± 0.017
AUROC	0.9052 ± 0.018	0.8900 ± 0.012	0.9000 ± 0.015	0.9957 ± 0.014

Table 4. Average classification performance (over 25 independent runs) on test MRI data using transfer learning with AlexNet

Model	Multiclass AD, EMCI, NC	Binary AD vs EMCI	Binary EMCI vs NC	Binary AD vs NC
Datasets	MRI	MRI	MRI	MRI
Acc	0.7200 ± 0.035	0.7463 ± 0.030	0.8500 ± 0.018	0.8600 ± 0.025
Precision	0.7050 ± 0.024	0.7376 ± 0.029	0.8600 ± 0.015	0.8600 ± 0.028
Recall	0.7268 ± 0.027	0.7525 ± 0.025	0.8431 ± 0.016	0.8600 ± 0.030
F-score	0.7157 ± 0.027	0.7450 ± 0.025	0.8515 ± 0.019	0.8600 ± 0.025
Specificity	0.7292 ± 0.030	0.7738 ± 0.023	0.8571 ± 0.025	0.8163 ± 0.031
MCC	0.4404 ± 0.025	0.4943 ± 0.020	0.7001 ± 0.020	0.7218 ± 0.030
AUROC	0.8550 ± 0.016	0.8900 ± 0.014	0.9200 ± 0.011	0.9600 ± 0.010

Figure 4 shows that the highest MCC score for the diagnosis of AD vs NC (0.98) is achieved with DTI data, whilst almost equal detection of early brain changes (EMCI vs NC) is achieved with both types of data.

Comparing the performance of the two transfer learning models one can observe that the highest performance of 89.50% (0.98 of AUC, 0.89 of F-score, 0.79 of MCC) in the multiclassification task is achieved with VGG-16 on MRI data. The best results in the binary classification tasks are obtained by VGG-16 nets using MRI data: AD vs EMCI (78% of accuracy, 0.88 of AUC, 0.78 of F-score); EMCI vs NC (93% of accuracy, 0.98 of AUC, 0.93 of F-score). The AD vs NC task is diagnosed better by transfer learning with the VGG-16 classifier when DTI data are used. AlexNet-based transfer learning also performs well on DTI data.

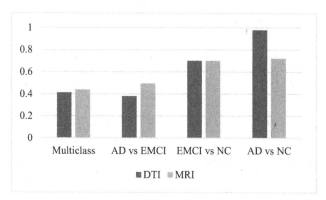

Fig. 4. AlexNet-based transfer learning MCC score in multiclass and binary classification of DTI and MRI data.

It is worth noticing that the time required for training and testing differs significantly between the AlexNet-based and the VGG-based transfer learning architectures (the same commodity hardware was used for all experiments as described in Sect. 3). AlexNet-based transfer learning required approximately 1.3 h for multiclassification and 0.85 h for binary classification, whilst transfer learning with the VGG-16 took 15.7 h and 9.7 h respectively.

6   Discussion and Conclusions

Deep transfer learning is a promising technique for the detection of cognitive decline when MRI data are used, as demonstrated by the experimental study. Nevertheless, DTI data appear to give an advantage to deep transfer learning when used for the diagnosis of Alzheimer's disease.

The performance of the classifiers used in the research indicates the advantage of the VGG-16-based models over the AlexNet ones using the transfer learning approach. For example, MCC coefficients are 0.995 (DTI set) and 0.871 (MRI set) for the diagnosis of AD with VGG-16 in the binary case, and 0.6899 (DTI set) and 0.7922 (MRI set) in the multiclass case. Also, MCC score with retrained AlexNet are as follows: 0.9799 (DTI set) and 0.7218 (MRI set) for the diagnosis of AD, and 0.7008 (DTI set) and 0.7001 (MRI set) for the diagnosis of MCI. However, the advantage of the VGG-16 comes at a price since these models take 8 to 15 times longer to train and test than AlexNet. The exploration of the additional CNN architectures, pretrained or created from scratch, can benefit further diagnosis of cognitive decline. From a medical perspective, the findings align with previous research that showed degeneration of the white matter of the brain is connected to and correlated with gray matter atrophy in cases of Alzheimer's disease. Axons of neurons can be affected earlier than the neurons themselves and can symbolize the early onset of the disease. DTI can detect these changes quicker than MRI and become the method of choice in the early diagnosis of Alzheimer's forms of dementia. The white matter in patients with MCI is affected significantly less. Thus, in the diagnosis of MCI and the transformation of some of its forms to AD, MRI technologies help computational models perform better compared to DTI. This can be explained by the fact that cognitive decline in the case of MCI might have different morphological grounds when the destructive process does not involve white matter only. The nature of MCI is more complex and might have another, vascular reason, for amnestic and cognitive decline. Only 30% of MCI progress to AD.

Lastly, there is potential to extend this research by focusing on longitudinal studies inside image classes, based on the evaluation and analysis of the changes of WM tracts during the progression of dementia by means of transfer learning with Convolutional Neural Networks. In this context, exploring the use of additional biomarkers (features) can potentially enhance the diagnosis of MCI using Deep Learning methods.

A List of Abbreviations

Alzheimer's disease (AD)
Convolutional Neural Networks (CNNs)
Densely connected networks (DenseNet)
Diffusion Tensor Imaging (DTI)
Gray matter (GM)
Matthews Correlation Coefficient (MCC)
Mild Cognitive Impairment (MCI)
Network in Networks (NiN)
Normal Controls (NC)
Random forest (RF)
Rectified linear unit (ReLU)
Residual Networks (ResNet)
Support vector machines (SVM)
Support vector regression (SVR)
Visual Geometry Group (VGG)
White matter (WM)

Acknowledgement. Data collection for this work was funded by the Alzheimer's Disease Neuroimaging Initiative (ADNI) (National Institutes of Health Grant U01 AG024904) and DOD ADNI (Department of Defense award number W81XWH-12-2-0012). The research presented in this paper was partially funded by a BEI School Award of Birkbeck College, University of London. For the purposes of open access, the author has applied a CC BY public copyright licence to any author accepted manuscript version arising from this submission.

References

1. Segato, A., Marzullo, A., Calimeri, F., De Momi, E.: Artificial Intelligence for brain diseases: a systematic review. APL Bioeng. **4**(4), 041503 (2020)
2. Soares, J., Marques, P., Alves, V., Sousa, N.: A hitchhiker's guide to diffusion tensor imaging. Front. Neurosci. **7**, 31 (2013)

3. Agosta, F., et al.: White matter damage in Alzheimer disease and its relationship to gray matter atrophy. Radiology **258**(3), 853–863 (2011)
4. Frings, L., et al.: Longitudinal grey and white matter changes in frontotemporal dementia and Alzheimer's disease. PLoS ONE **9**(3), e90814 (2014)
5. Mayo, C.D., Garcia-Barrera, M.A., Mazerolle, E.L., Ritchie, L.J., Fisk, J.D., Gawryluk, J.R.: Alzheimer's Disease Neuroimaging Initiative: Relationship between DTI metrics and cognitive function in Alzheimer's disease. Front. Aging Neurosci. **10**, 436 (2019)
6. Saha, A., Fadaiefard, P., Rabski, J.E., Sadeghian, A., Cusimano, M.D.: Machine learning applications using diffusion tensor imaging of human brain. a PubMed literature review. arXiv preprint arXiv:2012.10517 (2020)
7. Lundervold, A.S., Lundervold, A.: An overview of deep learning in medical imaging focusing on MRI. Zeitschrift für Medizinische Physik **29**(2), 102–127 (2019)
8. Alom, M.Z., et al.: The history began from AlexNet: a comprehensive survey on deep learning approaches. arXiv preprint arXiv:1803.01164 (2018)
9. Marzban, E.N., Eldeib, A.M., Yassine, I.A., Kadah, Y.M., Alzheimer's Disease Neurodegenerative Initiative: Alzheimer's disease diagnosis from diffusion tensor images using convolutional neural networks. PlOS One **15**(3), e0230409 (2020)
10. Nanni, L., Interlenghi, M., Brahnam, S., Salvatore, C., Papa, S., Nemni, R.: Alzheimer's Disease Neuroimaging Initiative: Comparison of transfer learning and conventional machine learning applied to structural brain MRI for the early diagnosis and prognosis of Alzheimer's disease. Front. Neurol. **1345** (2020)
11. Yagis, E., Citi, L., Diciotti, S., Marzi, C., Atnafu, S.W., De Herrera, A.G.S.: 3D convolutional neural networks for diagnosis of Alzheimer's disease via structural MRI. In: 2020 IEEE 33rd International Symposium on Computer-Based Medical Systems (CBMS), pp. 65–70. IEEE (2020)
12. Mehmood, A., et al.: A transfer learning approach for early diagnosis of Alzheimer's disease on MRI images. Neuroscience **460**, 43–52 (2021)
13. Mikołajczyk, A., Grochowski, M.: Data augmentation for improving deep learning in image classification problem. In: 2018 International Interdisciplinary PhD Workshop (IIPhDW), pp. 117–122. IEEE (2018)
14. Aderghal, K., Afdel, K., Benois-Pineau, J., Catheline, G.: Alzheimer's Disease Neuroimaging Initiative: Improving Alzheimer's stage categorization with Convolutional Neural Network using transfer learning and different magnetic resonance imaging modalities. Heliyon **6**(12), e05652 (2020)
15. LeCun, Y., Bengio, Y.: Convolutional networks for images, speech, and time series. In: The Handbook of Brain Theory and Neural Networks, vol. 3361, no.10 (1995)
16. Zhang, A., Lipton, Z.C., Li, M., Smola, A.J.: Dive into deep learning. arXiv preprint arXiv: 2106.11342 (2021)
17. Chollet, F.: Deep Learning with Python. Simon and Schuster (2021)
18. Nair, V., Hinton, G.E.: Rectified linear units improve restricted Boltzmann machines. In: ICML (2010)
19. Srivastava, N., Hinton, G., Krizhevsky, A., Sutskever, I., Salakhutdinov, R.: Dropout: a simple way to prevent neural networks from overfitting. J. Mach. Learn. Res. **15**(1), 1929–1958 (2014)
20. Krizhevsky, A., Sutskever, I., Hinton, G.E.: ImageNet classification with deep convolutional neural networks. In: Advances in Neural Information Processing Systems, vol. 25 (2012)
21. Simonyan, K., Zisserman, A.: Very deep convolutional networks for large-scale image recognition. arXiv preprint arXiv:1409.1556 (2014)

Improving Bacterial sRNA Identification By Combining Genomic Context and Sequence-Derived Features

Mohammad Sorkhian[1], Megha Nagari[1], Moustafa Elsisy[1],
and Lourdes Peña-Castillo[1,2(✉)] (iD)

[1] Department of Computer Science, Memorial University of Newfoundland,
St. John's, Canada
[2] Department of Biology, Memorial University of Newfoundland, St. John's, Canada
lourdes@mun.ca

Abstract. Bacterial small non-coding RNAs (sRNAs) are ubiquitous
regulatory RNAs involved in controlling several cellular processes by
targeting multiple mRNAs. The large diversity of sRNAs in terms of
their length, sequence, and function poses a challenge for computational
sRNA prediction. There are several bacterial sRNA prediction tools. Most
of them use sequence-derived features or rely on phylogenetic conservation. Recently, a new sRNA predictor (sRNARanking) showed that
using genomic context features outperformed methods based on sequence-
derived features. Here we comparatively assessed the effect of using
sequence-derived features together with genomic context features for computational sRNA prediction and generated a new model sRNARanking
v2 with increased predictive performance in terms of the area under the
precision-recall curve (AUPRC). sRNARanking v2 is available at:
https://github.com/BioinformaticsLabAtMUN/sRNARanking.

Keywords: Bacterial sRNA · Bioinformatics · Machine learning

1 Background

Bacterial small non-coding RNAs (sRNAs) are ubiquitous regulators of gene
expression, mostly acting by antisense mechanisms on multiple target mRNAs
and, as a result of this, they create complex regulatory networks [10]. Usually,
putative sRNAs are identified using RNA sequencing (RNA-seq) technologies
and their existence is validated by Northern blot analysis. Computational tools
for the identification of sRNAs can aid in the filtering of false sRNAs detected by
RNA-seq. Due to sRNAs' diversity in length and sequence [10], computational
sRNA identification remains a challenging task even though the first tools to
tackle this problem were developed decades ago [2,16].

There are two main approaches for sRNA identification: a) comparative-
genomics-based approaches which identifies sRNAs based on sequence similarity between putative sRNAs and known sRNAs (e.g., [12,21]); and b)

sequence-derived methods which compute features (such as k-mer frequencies, and free energy of secondary structure) from the sRNA sequence and use these features for classification (e.g., [1,3,23]). sRNAs often show large sequence differences between species or are present in only one species (i.e., they are species-specific) [24]. The proportion of species-specific sRNAs found per bacterium varies from one-fifth to nearly four-fifths [4,7,8], and comparative genomics approaches fail to identify these sRNAs. A recently developed approach, sRNARanking [6], proposed a third approach: the use of genomic context features for sRNA identification. Genomic context features encode the distance of a putative sRNA to other annotated genomic entities such as promoters, terminators, and open reading frames (ORFs).

In 2011, Lu et al. [16] assessed four comparative-genomics-based tools to identify sRNAs and found that on average their recall was between 20% to 49% with precisions of 6% to 12%. In 2014, Arnedo et al [1] selected seven existing methods to identify sRNAs, aggregated their predictions by applying union and intersection operations, optimized these aggregations to maximize specificity and sensitivity (recall), and combined the optimal aggregations by a majority voting strategy. Arnedo et al's proposed method (sRNA_OS) outperformed in terms of specificity and sensitivity the seven individual methods they used and the methods evaluated by Lu et al. Barman et al. [3] proposed a support vector machine (SVM) classifier using trinucleotide frequencies as features for sRNA identification. Their method outperformed Arnedo et al's method in terms of sensitivity, specificity, and accuracy. Tang et al. [23] investigated a variety of sequence-derived features and proposed two ensemble learning classifiers for sRNA identification. Their method has comparable predictive performance to that of Barman et al. In 2019, we proposed sRNARanking which is a random forest classifier using genomic context features for sRNA identification. More recently, Kumar et al. [14] presented Pred^{sRNA} which uses sequence-derived and secondary structure features to calculate a score. This score is used to discriminate between sRNAs and non-sRNA sequences.

sRNARanking substantially outperformed state-of-the-art methods, such as sRNA_OS and Barman et al's SVM; however, some questions remained such as a) whether sequence-derived features could approximate the performance obtained with genomic context features when the same training data was used, and b) whether a higher predictive performance could be obtained by combining sequence-derived and genomic context features. Here we answer these two questions. To do that we assessed the performance of random forest classifiers trained on the same data using different feature sets. Additionally we compared the performance of our best model, sRNARanking v2, on a multi-species dataset [16] and on a *Salmonella enterica serovar* Typhimurium LT2 (SLT2) dataset [3] with the performance of three other recent algorithms for the identification of sRNAs: PresRAT [14], sRNARanking [6], and Barman et al's SVM [3]. Our results show that combining both sequence-derived features and genomic context features generates the classifier with the best discriminative power in terms of area under the precision-recall curve (AUPRC).

2   Materials and Methods

2.1   Data

For hyper-parameter optimization and classifier selection, we use the data collected by [6]. These data contain sRNAs of five different bacterial species; namely, *Escherichia coli, Mycobacterium tuberculosis, Rhodobacter capsulatus, Salmonella enterica* and *Streptococcus pyogenes*. Positive instances of sRNAs are either verified by Northern blot analysis [8,13,15,18], homologous of known sRNAs [8], or listed in RegulonDB [22] as supported by literature with experimental evidence. Negative instances are randomly-selected genomic regions that do not overlap with the positive instances. To better reflect that out of all possible genomic regions only a few of them encode sRNAs, eight negative instances were generated for each positive instance. Negative instances were generated using bedtools [20] in three steps: 1) random genomic regions matching the length of the positive instances were generated using bedtools shuffle, 2) random genomic regions overlapping positive instances were filtered out using bedtools intersect, and 3) the negative instances' sequences were extracted from the corresponding bacterial whole genome using bedtools getfasta. After filtering negative instances overlapping with positive ones there is a positive to negative instances ratio of roughly 1:7. In total, our training data contain 341 positive instances and 2,462 negative instances for a total of 2,803 instances. As we have an imbalanced data set, we use performance metrics such as the Area Under the Precision-Recall Curve (AUPRC) which are suitable for imbalanced data.

For validating the performance of the classifiers, we use a multiple species dataset provided in [16] (henceforth referred to as Lu's data) and a *Salmonella enterica serovar* Typhimurium LT2 dataset provided in [3] (henceforth referred to as SLT2 data). Positive instances of Lu's data are provided in Supplementary Table S1 of [16] and positive instances of SLT2 data are in Table S6 of [3]. Negative instances were randomly-selected genomic regions from the corresponding genome that do not overlap with the positive instances (generated as described above). Using genomic regions not known to encode an sRNA instead of using artificial sequences gives a more conservative estimate of a classifier's performance and reflects more closely on how the model will be used by biologists. Eppenhof and Peña-Castillo showed that the performance of classifiers trained on artificial sequences substantially drops when used to distinguish sRNA sequences from real genomic sequences [6]. For Lu's data, there were three negative instances for each of the 754 positive instances, for a total of 3,309 instances. For the STL2 dataset, there were 10 negative instances for each of the 182 positive instances, for a total of 1986 instances. The intersection between the training data and the validation data is the empty set.

2.2   Feature Sets

Tang et al. [23] compared 17 sequence-derived feature sets for sRNA identification. The type of features evaluated by Tang et al. were spectrum profiles

(k-mer frequencies), mismatch profiles, reverse-complement k-mer profiles, and pseudo nucleotide composition. k-mer frequencies indicate the proportion of each k-mer relative to all k-mers present in a sequence, where k indicates the number of nucleotides considered; e.g., 3-mer indicates tri-nucleotides. Mismatch profiles are similar to k-mer frequencies but allow up to a certain number of mismatches m ($m < k$) [23]. Pseudo nucleotide composition includes 2-mer and 3-mer frequencies together with their physicochemical properties [23]. Mismatch profiles and pseudo nucleotide composition have parameters that need to be optimized. Additionally, Tang et al. found that random forest classifiers trained with these two types of features (mismatch profiles and pseudo nucleotide composition) do not outperform random forests trained with spectrum profiles or reverse-complement k-mer profiles in terms of the area under the Receiver Operating Characteristic curve (AUROC) [23]. Thus, we decided not to use mismatch profiles and pseudo nucleotide composition. Among the k-mer frequencies and reverse-complement k-mer frequencies, Tang et al's results show that the feature sets that generate the classifiers with the highest AUROC are tetra-nucleotide frequencies (4-spectrum profile) and reverse-complement tetra-nucleotide frequencies (i.e., the frequency of each tetra-nucleotide and its reverse complement are added together) [23]. Here we assessed these two sets of features: tetra-nucleotide frequencies (4-spectrum profile) and reverse-complement tetra-nucleotide frequencies. There are 256 tetra-nucleotide frequencies and 136 reverse-complement tetra-nucleotide frequencies.

As in [6], the seven genomic context features used are:

1. free energy of the sRNA predicted secondary structure,
2. distance to the closest predicted upstream promoter site in the range of [–1000, 0] nucleotides (if no promoter is predicted within this distance range a value of –1000 is used),
3. distance to the closest downstream predicted Rho-independent terminator in the range of [0, 1000] nucleotides (if no terminator is predicted within this distance range a value of 1000 is used),
4. distance to the closest left ORF, which is in the range of $(-\infty, 0]$ nucleotides,
5. a Boolean value (0 or 1) indicating whether the sRNA is transcribed on the same strand as its left ORF,
6. distance to the closest right ORF, which is in the range of $[0, +\infty)$ nucleotides (nts), and
7. a Boolean value indicating whether the sRNA is transcribed on the same strand as its right ORF.

A "left" ORF is an annotated ORF located at the 5' end of a genomic sequence on the forward strand or located at the 3' end of a genomic sequence on the reverse strand. A "right" ORF is an annotated ORF located at the 3' end of a genomic sequence on the forward strand or located at the 5' end of a genomic sequence on the reverse strand. Promoters are predicted using Promotech [5] and terminators are predicted using TransTermHP [11].

We generated a model using each of the following feature sets: 1) tetra-nucleotide frequencies, 2) reverse-complement tetra-nucleotide frequencies, 3) genome context, 4) tetra-nucleotide frequencies plus genome context, and 5) reverse-complement tetra-nucleotide frequencies plus genome context. To compute these features per bacterium, we used sRNACharP [6].

2.3   Model Generation

Eppenhof and Peña-Castillo found that random forest outperformed adaptive boosting, gradient boosting, logistic regression, and multilayer perceptron in the task of identifying bacterial sRNAs [6]. Thus, we decided to use random forest in this work.

Using grid-search stratified cross-validation on the training data, we optimized the number of features to consider at each split node ($max\_features$) and the total number of trees in the forest ($n\_estimators$). The values considered for $max\_features$ in the grid were 'log2', 'sqrt' and '0.33'; while the values considered for $n\_estimators$ were 100, 300, 500, 600, and 700. For all feature sets, the optimal $max\_features$ was the square root ('sqrt') of the total number of available features and the optimal $n\_estimators$ was 500. All other parameters were set to their default value.

Using the optimal values for $max\_features$ and $n\_estimators$, we evaluated the performance of each model using repeated 10-fold stratified cross-validation. The number of repeats was five; i.e., each model was trained and tested 50 times. The model's hyper-parameters with the highest mean average precision (which approximates AUPRC) were used to generate our final model using all the training data. All programs were run on Python 3.10.4 using the Python libraries scikit-learn [19] version 1.0.2, pandas [17] version 1.4.2 and numpy [9] version 1.22.3.

2.4   Comparative Assessment

AUPRC of our final model on Lu's and SLT2 data was compared with that of sRNARanking [6], PresRAT [14] and Barman et al's SVM [3]. To generate predictions with Barman et al's SVM method, we used the R code and proposed best model provided by Barman et al's to calculate the input features and obtain the predictions. To generate PresRAT predictions, we downloaded the Pred^{sRNA} software available at http://www.hpppi.iicb.res.in/presrat/Download.html. PresRAT was unable to make a prediction for sequences smaller than 30 nucleotides and sequences longer than 500 nucleotides. For these sequences (174 and 124 in Lu's and SLT2 data, respectively) we set the predicted probability to be a bona fide sRNA to 0.5. To obtain sRNARanking predictions, we used the R script provided by Eppenhof et al. [6]. All three methods (sRNARanking, Barman et al's SVM, and PresRat) were used as generated and provided by their original authors (i.e., these methods were not re-trained on our training data).

Table 1. Cross-validation average-precision. Mean average-precision was obtained using five repetions of 10-fold stratified cross-validation (i.e., 50 executions per feature set). Results are rounded up to two decimal places. GC refers to genomic context. Highest mean average-precision are highlighted in bold.

Model	Average AUPRC ± standard deviation
Tetra-nucleotides	0.49 ± 0.08
Reverse-complement tetra-nucleotide frequencies	0.46 ± 0.07
GC	0.71 ± 0.07
GC plus tetra-nucleotide frequencies	**0.78 ± 0.08**
GC plus reverse-complement tetra-nucleotide frequencies	**0.78 ± 0.08**

3   Results and Discussion

3.1   Model Selection

The two models with the highest cross-validation average precision are those generated with tetra-nucleotide frequencies plus genomic context features, and reverse-complement tetra-nucleotide frequencies plus genomic context features (Table 1). Using only sequence-derived features generates the models with the lowest average precision. We selected as our final model the one generated with the reverse-complement tetra-nucleotide frequencies plus genomic context features (henceforth referred to as sRNARanking v2), as it has fewer features than the model generated with the tetra-nucleotide frequencies plus genomic context features (136 vs 256).

The final random forest classifier was trained on the whole training data with the following hyper-parameters: $max\_features$ (number of features to consider at each split) equal to 'sqrt' (i.e., the number of features to consider is the square root of the total number of available features), and $n\_estimators$ (number of tree in the random forest) equal to 500. All other hyper-parameters were left to the default value of the sklearn.ensemble.RandomForestClassifier function.

3.2   Variable Importance Analysis

We looked at which attributes are more important in our final random forest model by calculating the average decrease in accuracy caused by randomly permuting each feature. To do this we used the scikit-learn function permutation_importance. Each feature was permuted ten times.

The five most important variables are all genomic context features; namely, distance to the closest ORFs, distance to the closest terminator, distance to the closest promoter, and free energy of the sRNA predicted secondary structure (Fig. 1). The average decrease in accuracy of the sequence-derived features varies from zero to 0.0006 with a median value of zero and a mean value of

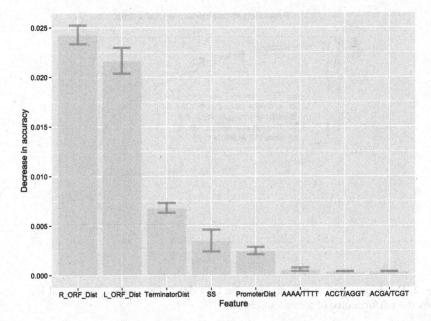

Fig. 1. Variable importance by decrease in accuracy obtained by permuting each feature ten times. The most important features are distances to the closest ORFs (R_ORF_Dist and L_ORF_Dist), distance to the closest downstream terminator (TerminatorDist), free energy of the sRNA predicted secondary structure (SS), and distance to the closest upstream promoter (PromoterDist).

1.076×10^{-05}. This indicates that sequence-derived features have a small positive effect on the accuracy of sRNARanking v2. As genomic context features are the most important features, classification performance might be further improved by adding other relevant genomic context features. As future work, we will explore whether other genomic context features can be added to improve the predictive performance of computational tools for the identification of sRNAs.

3.3   Comparative Assessment

We compared the performance of sRNARanking v2 in terms of AUPRC with the AUPRC of PresRAT [14], Barman's SVM model [3] and sRNARanking [6] on two data sets: Lu's (consisting of sequences from 14 bacteria) and SLT2. No sequence on these two data sets was in the training data. sRNARanking v2 substantially outperformed PresRAT and Barman's SVM model on both data sets (Figs. 2 and 3). Additionally, sRNARanking v2 had an improvement of 0.04 and 0.13 in AUPRC on Lu's data and SLT2 data, respectively, with respect to sRNARanking's AUPRC (Figs. 2 and 3). This indicates that indeed sequence-derived features such as reverse-complement tetra-nucleotide frequencies improve discriminative power on top of genomic context features.

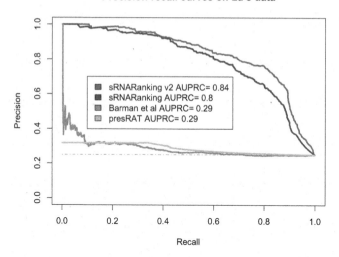

Fig. 2. Precision recall curves obtained on Lu's data. The dashed horizontal line indicates the performance of a random classifier.

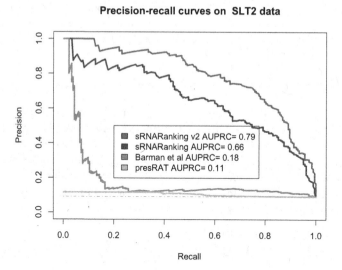

Fig. 3. Precision recall curves obtained on SLT2 data. The dashed horizontal line indicates the performance of a random classifier.

We observed that the positive effect of adding sequences-derived features was more marked in the SLT2 data than in Lu's data. We hypothesized that adding sequence-derived features had a larger beneficial effect on SLT2 data because sequences on this data set have a more similar reverse-complement tetra-nucleotide frequency distribution with those in the training data than some of

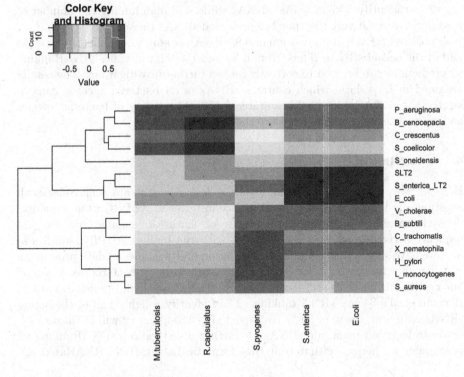

Fig. 4. Pairwise Spearman's correlation coefficients between the average reverse-complement tetra-nucleotide frequencies of bacteria in the validation data (rows) with bacteria in the training data (columns). Value in the color key refers to the Spearman's correlation value.

the bacteria in Lu's data. To explore this hypothesis, we obtained the average reverse-complement tetra-nucleotide frequencies per bacterium and calculated Spearman's correlation coefficient between the reverse-complement tetra-nucleotide frequencies per each pair of bacteria on the training data and the validation data (Lu's and SLT2 data sets) (Fig. 4). Positive pair-wise Spearman's correlation values indicate that those two bacteria have similar reverse-complement tetra-nucleotide frequency distribution. Negative pair-wise Spearman's correlation values, for example, the one between *L. monocytogenes* and *M. tuberculosis* (−0.48), indicate an inverse relationship. The bacterium with the highest average Spearman's correlation value (0.557) with those in the training data was *Salmonella enterica serovar* Typhimurium LT2 (SLT2). The average Spearman's correlation value of the bacteria in Lu's data with those in training data goes from 0.028 (*Staphylococcus aureus subsp. aureus* N315) to 0.500 (*E. coli*). This suggests that adding more diverse bacteria to the training data might further improve the predictive performance of our model.

At a recall of 0.8, sRNARanking v2 achieves a precision of 0.76 and 0.63 in Lu's and SLT2 data sets, respectively (Figs. 2 and 3). In other words, sRNARanking v2 can identify 80% of actual sRNAs while still maintaining the number of false positives relatively low (per 100 predicted sRNAs there will be less than 37 false positives). Even at a precision of 0.9, sRNARanking v2 can identify close to half of the actual sRNAs. Thus, from a biologist's point of view, sRNARanking v2 predictions can be used to correctly inform further investigations. The results obtained in Lu's data, which contains sRNAs of 14 bacterial species, demonstrate that sRNARanking v2 is suitable for a wide variety of bacterial species and generalizes beyond the five bacteria used for training.

4   Conclusion

Here we have shown that genomic context features generate a classifier with substantially higher predictive performance in terms of AUPRC than sequence-derived features. Combining both types of features, genomic context and sequence-derived, generates the classifier with the highest AUPRC. Additionally, we have shown that training a classifier on multiple-species data produces a classifier that can better identify sRNAs of a wide range of bacteria (as shown by our results in Lu's data set) than classifiers trained on single species data (e.g., Barman et al's SVM). sRNARanking v2 can identify roughly half of the actual sRNAs with a precision of 90%. We expect sRNARanking v2 will facilitate biologists to focus on bona fide sRNAs for further investigation. sRNARanking v2 is available at: https://github.com/BioinformaticsLabAtMUN/sRNARanking.

Acknowledgments. This research was partially supported by a grant from the Natural Sciences and Engineering Research Council of Canada (NSERC) to L.P.-C. (Grant number RGPIN: 2019-05247). This research was enabled in part by support provided by ACENET (www.ace-net.ca/) and Compute Canada (www.computecanada.ca).

References

1. Arnedo, J., Romero-Zaliz, R., Zwir, I., Del Val, C.: A multiobjective method for robust identification of bacterial small non-coding RNAs. Bioinformatics **30**(20), 2875–82 (2014). https://doi.org/10.1093/bioinformatics/btu398
2. Backofen, R., Hess, W.R.: Computational prediction of sRNAs and their targets in bacteria. RNA Biol. **7**(1), 33–42 (2010)
3. Barman, R.K., Mukhopadhyay, A., Das, S.: An improved method for identification of small non-coding RNAs in bacteria using support vector machine. Sci. Rep. **7**, 46070 (2017). https://doi.org/10.1038/srep46070
4. Broach, W.H., Weiss, A., Shaw, L.N.: Transcriptomic analysis of staphylococcal sRNAs: insights into species-specific adaption and the evolution of pathogenesis. Microb. Genom. **2**(7), e000065 (2016). https://doi.org/10.1099/mgen.0.000065
5. Chevez-Guardado, R., Peña-Castillo, L.: Promotech: a general tool for bacterial promoter recognition. Genome. Biol. **22**(1), 318 (11 2021). https://doi.org/10.1186/s13059-021-02514-9
6. Eppenhof, E.J.J., Peña-Castillo, L.: Prioritizing bona fide bacterial small RNAs with machine learning classifiers. PeerJ **7**, e6304 (2019). https://doi.org/10.7717/peerj.6304

7. Gómez-Lozano, M., Marvig, R.L., Molina-Santiago, C., Tribelli, P.M., Ramos, J.L., Molin, S.: Diversity of small RNAs expressed in Pseudomonas species. Environ. Microbiol. Rep. **7**(2), 227–36 (2015). https://doi.org/10.1111/1758-2229.12233. Apr

8. Grüll, M.P., Peña-Castillo, L., Mulligan, M.E., Lang, A.S.: Genome-wide identification and characterization of small RNAs in Rhodobacter capsulatus and identification of small RNAs affected by loss of the response regulator CtrA. RNA Biol. 1–12 (2017). https://doi.org/10.1080/15476286.2017.1306175

9. Harris, C.R., et al.: Array programming with NumPy. Nature **585**(7825), 357–362 (2020) https://doi.org/10.1038/s41586-020-2649-2

10. Hör, J., Gorski, S.A., Vogel, J.: Bacterial RNA biology on a genome scale. Mol. Cell **70**(5), 785–799 (2018). https://doi.org/10.1016/j.molcel.2017.12.023

11. Kingsford, C.L., Ayanbule, K., Salzberg, S.L.: Rapid, accurate, computational discovery of Rho-independent transcription terminators illuminates their relationship to DNA uptake. Genome. Biol. **8**(2), R22 (2007). https://doi.org/10.1186/gb-2007-8-2-r22

12. Klein, R.J., Misulovin, Z., Eddy, S.R.: Noncoding RNA genes identified in AT-rich hyperthermophiles. Proc. Natl. Acad. Sci. USA **99**(11), 7542–7 (2002). https://doi.org/10.1073/pnas.112063799

13. Kröger, C., et al.: The transcriptional landscape and small RNAs of Salmonella enterica serovar typhimurium. Proc. Natl. Acad. Sci. USA **109**(20), E1277-86 (2012). https://doi.org/10.1073/pnas.1201061109

14. Kumar, K., Chakraborty, A., Chakrabarti, S.: PresRAT: a server for identification of bacterial small-RNA sequences and their targets with probable binding region. RNA Biol. **18**(8), 1152–1159 (2021). https://doi.org/10.1080/15476286.2020.1836455

15. Le Rhun, A., Beer, Y.Y., Reimegård, J., Chylinski, K., Charpentier, E.: RNA sequencing uncovers antisense RNAs and novel small RNAs in Streptococcus pyogenes. RNA Biol. **13**(2), 177–95 (2016). https://doi.org/10.1080/15476286.2015.1110674

16. Lu, X., Goodrich-Blair, H., Tjaden, B.: Assessing computational tools for the discovery of small RNA genes in bacteria. RNA **17**(9), 1635–47 (2011). https://doi.org/10.1261/rna.2689811. Sep

17. McKinney, W.: Data structures for statistical computing in python. In: van der Walt, S., Millman, J. (eds.) Proceedings of the 9th Python in Science Conference, pp. 56–61 (2010). https://doi.org/10.25080/Majora-92bf1922-00a

18. Miotto, P., et al.: Genome-wide discovery of small RNAs in Mycobacterium tuberculosis. PLoS One **7**(12), e51950 (2012). https://doi.org/10.1371/journal.pone.0051950

19. Pedregosa, F., et al.: Scikit-learn: Machine learning in Python. J. Mach. Learn. Res. **12**, 2825–2830 (2011)

20. Quinlan, A.R., Hall, I.M.: BEDTools: a flexible suite of utilities for comparing genomic features. Bioinformatics **26**(6), 841–2 (2010). https://doi.org/10.1093/bioinformatics/btq033. Mar

21. Rivas, E., Eddy, S.R.: Noncoding RNA gene detection using comparative sequence analysis. BMC Bioinform. **2**, 8 (2001). https://doi.org/10.1186/1471-2105-2-8

22. Santos-Zavaleta, A., et al.: RegulonDB v 10.5: tackling challenges to unify classic and high throughput knowledge of gene regulation in E. coli K-12. Nucleic Acids Res. **47**(D1), D212–D220 (2019). https://doi.org/10.1093/nar/gky1077

23. Tang, G., Shi, J., Wu, W., Yue, X., Zhang, W.: Sequence-based bacterial small RNAs prediction using ensemble learning strategies. BMC Bioinform. **19**(Suppl 20), 503 (2018). https://doi.org/10.1186/s12859-018-2535-1. Dec

24. Wagner, E.G.H., Romby, P.: Small RNAs in bacteria and archaea: who they are, what they do, and how they do it. Adv. Genet. **90**, 133–208 (2015). https://doi.org/10.1016/bs.adgen.2015.05.001

High-Dimensional Multi-trait GWAS By Reverse Prediction of Genotypes Using Machine Learning Methods

Muhammad Ammar Malik$^{(\boxtimes)}$ (iD), Adriaan-Alexander Ludl(iD),
and Tom Michoel(iD)

Computational Biology Unit, Department of Informatics, University of Bergen,
PO Box 7803, 5057 Bergen, Norway
{muhammad.malik,adriaan.ludl,tom.michoel}@uib.no

Abstract. Multi-trait genome-wide association studies (GWAS) use multi-variate statistical methods to identify associations between genetic variants and multiple correlated traits simultaneously, and have higher statistical power than independent univariate analyses of traits. Reverse regression, where genotypes of genetic variants are regressed on multiple traits simultaneously, has emerged as a promising approach to perform multi-trait GWAS in high-dimensional settings where the number of traits exceeds the number of samples. We analyzed different machine learning methods (ridge regression, naive Bayes/independent univariate, random forests and support vector machines) for reverse regression in multi-trait GWAS, using genotypes, gene expression data and ground-truth transcriptional regulatory networks from the DREAM5 SysGen Challenge and from a cross between two yeast strains to evaluate methods. We found that genotype prediction performance, in terms of root mean squared error (RMSE), allowed to distinguish between genomic regions with high and low transcriptional activity. Moreover, model feature coefficients correlated with the strength of association between variants and individual traits, and were predictive of true trans acting expression quantitative trait loci (trans-eQTL) target genes, with complementary findings across methods.

Keywords: Genome-wide association studies · Machine learning ·
Multi-trait GWAS · Gene expression · Genotype prediction

1 Background

Genome-wide association studies (GWAS) aim to find statistical associations between genetic variants and traits of interest using data from a large number of individuals [1,2]. When multiple correlated traits are studied simultaneously, joint, multi-trait approaches can be more advantageous than studying the

---

Supplementary Information The online version contains supplementary material available at https://doi.org/10.1007/978-3-031-20837-9_7.

traits individually, due to increased power from taking into account cross-trait covariances and reduced multiple-testing burden by performing a single test for association to a set of traits [3–5].

The most commonly used multi-trait GWAS approaches are based on a multivariate analysis of variance (MANOVA) or canonical correlation analysis (CCA) [3]. However, these are applicable only to studies where the number of traits is relatively small, especially in comparison to the number of samples. When analyzing the effects of genetic variants on molecular traits (gene or protein expression levels, metabolite concentrations) or imaging features, we have to deal with a large number, often an order of magnitude or more greater than the sample size, of correlated traits simultaneously. For such studies, the standard procedure is still to conduct univariate linear regression or ANOVA tests for each genetic variant against each trait separately. While efficient algorithms exist to undertake this task [6–8], the massive multiple-testing problem results in a significant loss of statistical power.

An alternative approach to multi-trait GWAS has been to reverse the functional relation between genotypes and traits, and fit a multivariate regression model that predicts genotypes from multiple traits simultaneously, instead of the usual approach to regress traits on genotypes. The first study to do this explicitly used logistic regression and showed a significant increase in power compared to univariate methods, without being dependent on assuming normally distributed genotypes like MANOVA or CCA [9]. Although the method as presented in [9] is still only valid when the number of traits is small, extending multivariate regression methods to high-dimensional settings is straightforward. Thus a recent study used L2-regularized linear regression of single nucleotide polymorphisms (SNPs) on gene expression traits to identify trans-acting expression quantitative trait loci (trans-eQTLs), and showed that this approach aggregates evidence from many small trans-effects while being unaffected by strong expression correlations [10]. In a very different application domain, regularized regression of SNP genotypes on longitudinal image phenotypes was used to identify time-dependent genetic associations with imaging phenotypes [11].

Despite these advances, several limitations and open questions remain unanswered in high-dimensional GWAS. Firstly, linear models search for the linear combination of traits that is most strongly associated to the genetic variant, but there is no a priori biological reason why only linear combinations should be considered. Secondly, while L2-regularization allows to deal with high-dimensional traits, it does not address the problem of variable selection. For instance, in the case of gene expression, we expect that trans-eQTLs are potentially associated with many, but not all genes. Indeed, in [10] a secondary set of univariate tests is carried out to select genes associated to trans-eQTLs identified by the initial multi-variate regression. Thirdly, a systematic biological validation and comparison of the available methods is lacking.

Here we address these questions by considering a wider range of machine learning methods (in particular, random forests (RF) and support vector machines (SVM)) for reverse genotype prediction from gene expression traits.

Hypothesizing that true trans-eQTL associations are mediated by transcription regulatory networks, we use simulated data from the DREAM5 Systems Genetics Challenge, and real data from 1,012 segregants of a cross between two budding yeast strains [12] together with the YEASTRACT database of known transcriptional interactions [13], to validate and compare these methods against univariate linear correlation (naive bayes) and ridge regression.

2   Methods

2.1   Reverse Genotype and Trans-eQTL Prediction

For genotype prediction using machine learning models, the expression values were treated as explanatory variables whereas the genotype value of a variant was treated as a response variable. The prediction performance was measured by computing the root mean squared error (RMSE) between the predicted and the actual genotype value of variants.

Trans-eQTL target prediction was done using weights assigned to the features by the machine learning methods: feature importance in case of random forest regression (RFR), and coefficients for support vector regression (SVR) and ridge regression (RR). We computed the area under the receiver operating characteristic (AUROC) curve to measure prediction performance by comparing the weights against the true targets in the ground truth for each variant.

2.2   Datasets

Simulated Data. The simulated data for our experiments was obtained from DREAM5 Systems Genetic Challenge A[1], generated by the SysGenSIM software [14]. The DREAM data consists of simulated genotype and transcriptome data of synthetic gene regulatory networks. The dataset consists of 15 sub-datasets, where 5 different networks are provided and for each network 100, 300 and 999 samples are simulated. Every sub-dataset contains 1000 genes. We used the networks with 999 samples only.

In the DREAM data, each genetic variant is associated to a unique causal gene that mediates its effect. We therefore defined ground-truth trans-eQTL targets for each variant as the causal gene's direct targets in the ground-truth network.

In the DREAM data 25% of the variants acted in *cis*, meaning they affected expression of their causal gene directly. The remaining 75% of the variants acted in *trans*. Since the identities of the *cis* and *trans* eQTLs are unknown, we computed the P-values of genotype-gene expression associations between matching variant-gene pairs using Pearson correlation and selected all genes with P-values less than 1/750 to identify cis-acting eQTLs.

---

[1] https://www.synapse.org/#!Synapse:syn2820440/wiki/.

Yeast Data. The yeast data used in this paper was obtained from [12]. The expression data contains expression values for 5,720 genes in 1,012 segregants. The genotype data consists of binary genotype values for 42,052 genetic markers in the same 1,012 segregants.

Batch and optical density (OD) effects, as given by the covariates provided in [12], were removed from the expression data using categorical regression, as implemented in the *statsmodels* python package. The expression data was then normalized to have zero mean and unit standard deviation.

To match variants to genes, we considered the list of genome-wide significant eQTLs provided by [12] whose confidence interval (of variable size) overlapped with an interval covering a gene plus 1,000 bp upstream and 200 bp downstream of the gene position. This resulted in a list of 2,884 genes and for each of these genes we defined its matching variant as the most strongly associated variant from the list.

Networks of known transciptional regulatory interactions in yeast (S. cerevisiae) were obtained from the YEASTRACT (Yeast Search for Transcriptional Regulators And Consensus Tracking) [13]. Regulation matrices were also obtained from YEASTRACT[2]. We retrieved the ground-truth matrix containing all reported interactions of the type *DNA binding and expression evidence*. Self regulation was removed from the ground-truths. The Ensembl database (release 83, December 2015) [15] was used to map gene names to their identifiers. After overlaying the ground-truth with the set of genes with matching cis-eQTL, a ground-truth network of 80 transcription factors (TFs) with matching cis-eQTL and 3,394 target genes was obtained.

The expression dataset was then filtered to contain only the genes present in the ground truth network, and ground-truth trans-eQTL sets for the 80 TF-associated cis-eQTL genetic variants were defined as direct targets of the corresponding TFs in the ground-truth network.

2.3 Experimental Settings

In all sets of experiments we used 5-fold cross-validation. Ridge Regression (RR), Random Forest Regression (RFR), Support Vector Regression (SVR), and Naive Bayes (NB) were implemented using the Python library *scikit-learn*. For RR, the regularization strength (α) was set to 100 and other parameters were set to their defaults. For RR and SVR, the default parameters were used. For NB, we used the Gaussian Naive Bayes from *scikit-learn* library. For trans-eQTL predictions, univariate linear correlation was also used to compare with the regression methods mentioned above.

Feature Selection. For each method we took the absolute values of the feature importances/coefficients, scaled so that their sum equals to 1, and sorted these in descending order. These scaled values represent the relative contribution of each

---

[2] http://www.yeastract.com/formregmatrix.php.

(A)   **(B)**

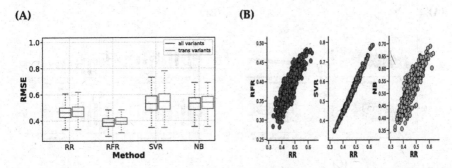

Fig. 1. RMSE values for genotype prediction on DREAM5 simulated data. (**A**) Box-plots show the distribution of the RMSE values for all variants (blue) and for trans-acting-only variants (red) for random forest regression (RFR), support vector regression (SVR), ridge regression (RR), and naive Bayes (NB). (**B**) Scatter plots show RMSE values of RFR, SVR, and NB vs RR for all variants. The data shown are for DREAM Network 1. The results for Network 2–5 are shown in Supp. Figs. S1–S4. (Color figure online)

feature to the prediction of each variant. We selected the top-scoring features which together contributed 50% of the feature weight sum.

2.4 Code and Supplementary Information

The scripts to reproduce the analysis and the supplementary information are available at https://github.com/michoel-lab/Reverse-Pred-GWAS.

3 Results

3.1 Reverse Genotype Prediction and Trans-EQTL Analysis in Simulated Data

In the DREAM5 Systems Genetics Challenge, binary genotypes and steady-state gene expression data for 1,000 genes were simulated for a population of 999 individuals, based on a gene network topology and the individuals' genotypes at a set of genome-wide DNA variants, using non-linear ordinary differential equations (ODEs) [14]. In the simulations, there was a one-to-one mapping between genetic variants and genes, such that the effects of each variant are mediated by exactly one causal gene. 25% of the variants acted in *cis*, meaning they affected expression of their causal gene, but not the value of any of the parameters in the ODE model. The remaining 75% of the variants acted in *trans*, meaning they did not affect expression of their causal gene, but did affect the transcription rate of the causal gene's targets in the network. Simulated data for five networks are available.

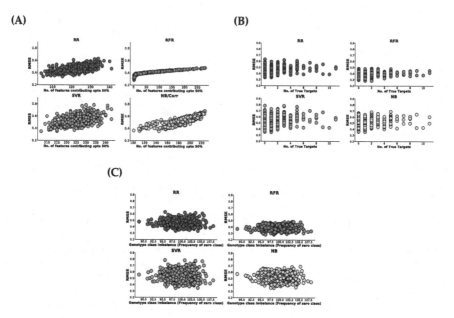

Fig. 2. Scatter plots of genotype RMSE values on DREAM5 simulated data against the number of selected model features (**A**), the number of true trans-eQTL targets in the ground-truth network (**B**), and the genotype class balance (frequency of the zero class) (**C**), for random forest regression (RFR), support vector regression (SVR), ridge regression (RR), and naive Bayes (NB). The data shown are for DREAM Network 1. The results for Network 2–5 are shown in Supp. Figs. S5–S8.

Genotype Prediction Accuracy Varies Across Genetic Variants. We trained models to predict the genotypes for variants whose causal gene had at least one target in the ground-truth network (covering between 491–644 genes depending on the network/dataset) using the expression data from all 1,000 genes as predictors, using Random Forest Regression (RFR), Support Vector Regression (SVR), Ridge Regression (RR) and Naive Bayes classification (NB). RMSE was then measured for each predicted variant in the test data. Mean performance across the five train-test folds is reported here.

RFR achieved the best prediction performance (lowest RMSE) overall (RMSE ∼0.3–0.5). RR achieved RMSEs in the range of ∼0.6–0.8. In contrast to ·RFR and RR, the RMSE varied widely for SVR and NB (∼0.3–0.9) (Fig. 1A). We did not observe a significant change in the distribution of RMSE values for all the variants versus keeping only *trans*-acting variants (Fig. 1A), i.e. *cis*-acting variants are not significantly easier to predict (by virtue of having a highly correlated *cis*-gene) than variants that only have *trans*-associated genes. While RMSE values are correlated between the methods (Fig. 1B), the correlation is imperfect (with the exception of SVR-RR), such that there is considerable variation in the RMSE-based ranking of variants between the methods.

Taken together these result show that prediction performance varies across genetic variants within each method (i.e. variants can be ranked according to their RMSE) and that RFR can be preferred over the others in terms of average prediction performance, but with considerable variation in relative performance across methods for individual variants.

Next, we compared the genotype prediction performance for the different methods with the number of features contributing upto 50% of the total sum of feature weights (cf. Methods). In general, variants that were more predictable had models with fewer features, and vice versa, irrespective of the prediction method used (Fig. 2A). On the other hand, we did not observe any significant relation between the prediction performance and the number of true targets in the ground truth network (Fig. 2B). We also tested whether RMSE was influenced by the genotype class imbalance. This was not the case for the regression-based methods used here (Fig. 2C).

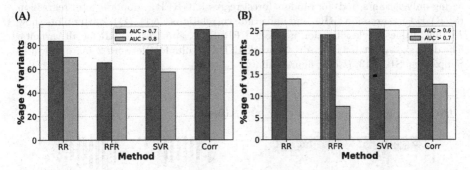

Fig. 3. A. Bar plots show the proportion of variants with trans-eQTL target prediction AUROC > 0.7 (blue) and > 0.8 (red) in DREAM5 simulated data. **B.** Bar plots show the number of variants with trans-eQTL target prediction AUROC ≥ 0.6 (blue) and ≥ 0.7 (red) in yeast data. Genes on the same chromosome were excluded as predictors for each SNP. (Color figure online)

Feature Importances are Predictive of True Trans-EQTL Associations.

To evaluate the ability of reverse genotype prediction methods to identify true trans-eQTL targets of a given variant, we defined true trans associations as direct target genes of a variant's causal gene in the ground-truth network and used feature importances/coefficients in the genotype prediction model to predict how likely a gene is to be a trans-eQTL of a given variant (see Methods). Performance was measured using the area under the receiver operating curve (AUROC).

For all methods, more than $\sim$55%, resp. $\sim$65% of variants with at least one trans-eQTL target in the ground-truth network had AUROC> 0.8, resp. 0.7, with univariate linear correlation and ridge regression performing somewhat better than random forest and SVR (Fig. 3A). Ridge regression and univariate correlation methods also had less variation in terms of AUROCs when compared with RFR and SVR, and no significant difference in terms of AUROC was observed when using all the variants versus using only *trans*-acting variants (Fig. 4A). Interestingly, the variants for which high AUROCs were obtained differed between RFR, RR and univariate correlation methods, whereas RR and SVR obtained nearly identical performance on all variants. (Fig. 4B).

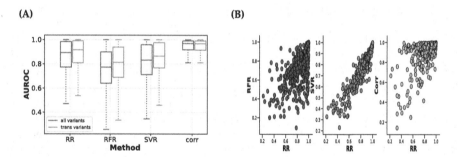

Fig. 4. Trans-eQTL target prediction performance on DREAM5 simulated data. (**A**) Boxplots show the distribution of AUROC values for all variants (blue) and for trans-acting-only variants (red) for random forest regression (RFR), support vector regression (SVR), ridge regression (RR), and univariate correlation (Corr). (**B**) Scatter plots show AUROC values of classification methods RFR, SVR, and Corr vs RR for all variants. The data shown are for DREAM Network 1. The results for Network 2–5 are shown in Supp. Figs. S10–S13. (Color figure online)

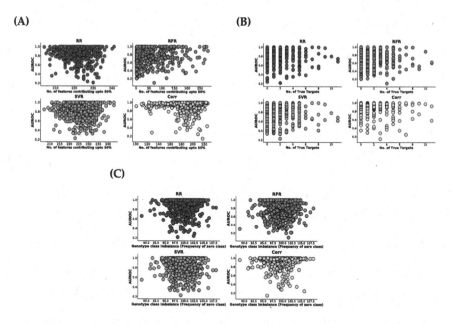

Fig. 5. Scatter plots of trans-eQTL target prediction performance (AUROC) on DREAM5 simulated data against the number of selected model features (**A**), the number of true trans-eQTL targets in the ground-truth network (**B**), and the genotype class balance (frequency of the zero class) (**C**), for random forest regression (RFR), support vector regression (SVR), ridge regression (RR), and univariate correlation/naive Bayes (NB). The data shown are for DREAM Network 1. The results for Network 2–5 are shown in Supp. Figs. S14–S17.

(A)   **(B)**

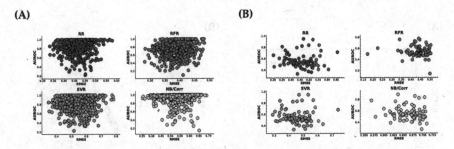

Fig. 6. Scatter plots show trans-eQTL prediction accuracy (AUROC) vs genotype prediction performance (RMSE) for random forest regression (RFR), support vector regression (SVR), ridge regression (RR), and univariate correlation/naive Bayes (NB/Corr). (**A**). For DREAM5 simulated data. (**B**). For yeast data where genes on the same chromosome were excluded as predictors for each SNP

When compared to potential explanatory factors, no significant relation was observed between AUROC values and number of selected model features (Fig. 5A), number of ground-truth targets (Fig. 5B), or the genotype class balance (Fig. 5C).

Genotype and Trans-EQTL Prediction Performance Do Not Correlate. Finally we tested whether genotype prediction accuracy can be used as a proxy for trans-eQTL prediction accuracy, that is, in the absence of ground-truth networks, can we use genotype prediction accuracy to filter variants whose model feature weights are indicative of true trans-eQTL targets? However, we did not observe any correlation between the genotype prediction performance and trans-eQTL target prediction performance for any of the methods (Fig. 6A)

3.2 Reverse Genotype Prediction and Trans-EQTL Analysis in Yeast

In the next set of experiments we repeated the same analysis on yeast dataset. Compared to the simulated data, the yeast data differs in two important aspects. First, ground-truth target information is available for a small set of transcription factors (TFs) only. Secondly, we have no knowledge of the causal gene(s) corresponding to each variant, and need to rely on a local *cis*-association between a variant and a TF to define a ground-truth set of trans-eQTL targets to a variant (cf. Methods).

Genotype Prediction Accuracy Varies Across Genetic Variants. Genotype prediction performance for the yeast data also varied across genetic variants. Similar to DREAM data RFR achieved lowest RMSE values in the yeast data as well (Fig. 7A). We tested whether prediction performance may be explained by local *cis*-associations by removing genes on the same chromosome as the

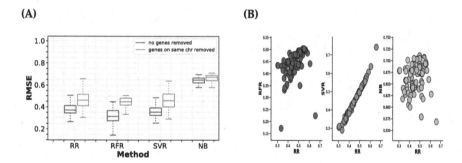

Fig. 7. Genotype prediction performance on yeast data. Genotype prediction performance on yeast data. (**A**) Boxplots show the distribution of the performance for all variants using all genes (blue) and excluding genes on the same chromosome as the variant (red) as predictors, for random forest regression (RFR), support vector regression (SVR), ridge regression (RR), and naive Bayes (NB). (**B**) Scatter plots show RMSE values of classification methods RFR, SVR, and NB vs RR for all variants. Genes on the same chromosome were excluded as predictors for each SNP. (Color figure online)

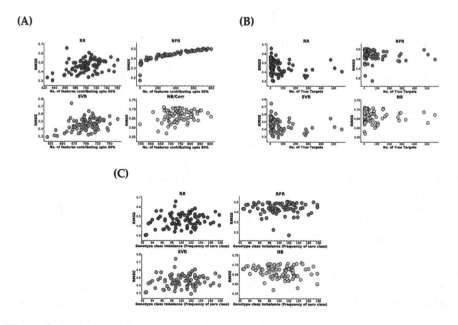

Fig. 8. Scatter plots of genotype prediction performance on yeast data against the number of selected model features (**A**), the number of true trans-eQTL targets in the ground-truth network (**B**) and the genotype class balance (frequency of the zero class) (**C**), for random forest regression (RFR), support vector regression (SVR), ridge regression (RR), and naive Bayes (NB). Genes on the same chromosome were excluded as predictors for each SNP.

test variant from the list of predictors. In this case we did observe that RMSE increased markedly (i.e. prediction performance decreased) when removing local genes, except for NB, and that after removing local genes, RR, RFR, and SVR have similar average prediction performance (Fig. 7A).

Correlations of RMSEs between methods showed a similar pattern as in the simulated data, with RR and SVR RMSEs being particularly strongly correlated (Fig. 7B).

As in the simulated data, genotype prediction performance decreased (i.e. RMSE increased) with increasing number of model features (Fig. 8A), but did not depend significantly on the number of true targets (Fig. 8B) or genotype class balance (Fig. 8C).

Next we tested whether feature importance weights were predictive of true trans-eQTL associations, defined as genes that were bound by and differentially expressed upon perturbation of a TF for which a given variant is a cis-eQTL (cf. Methods). In this case, feature importances were only modestly predictive, with 20–30%, resp. 10–15%, of TF cis-eQTLs obtaining AUROCs > 0.6, resp. > 0.7, and, as in the simulated data, there were fewer variants with high AUROC for RFR, compared to the other methods (Fig. 3B).

We confirmed that the distribution of AUROC values was not affected by removing genes on the same chromosome as a variant of interest from the list of predictors (Fig. 9A). Furthermore, the AUROC values showed no relation with the number of selected features, the number of true targets, or the genotype class balance (Fig. 10).

Although AUROCs generally correlated between methods (Fig. 9B), in line with the correlation of RMSE values, AUROC values tended to be systematically higher for SVR and RR compared to RFR and univariate correlations. Interestingly, univariate correlation and SVR share the same number of TF eQTLs with AUROC > 0.70 (10), only 5 were common and each method had five TFs not found by the other method (Supp. Figs. S22).

Genotype and Trans-EQTL Prediction Performance Do Not Correlate. Similar to the DREAM data we again observed poor correlation between genotype and trans-eQTL prediction performance (Fig. 6B).

Feature Selection in Random Forest Produces a Map of Transcriptional Hotspots. Transcriptional hotspots are regions of the genome associated with widespread changes in gene expression [12]. We learned prediction models for all 2,884 SNPs in the yeast genome that were associated with local changes in gene expression and plotted the RMSE for each predicted SNP against its genome position. RFR showed a wide variation in RMSE values for SNPs, across the whole genome, allowing to delineate genomic ranges with high and low regulatory activity. Whereas RR and SVR showed much less variation, and did not allow to separate high and low activity regions on most chromosomes (Fig. 11). Interestingly, the regions detected by RFR overlapped only partially with traditional hotspot maps based on univariate correlations (Supp. Fig. S23),

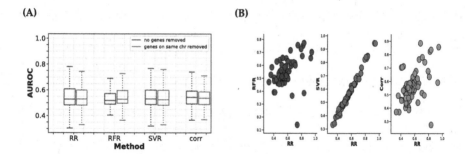

Fig. 9. Trans-eQTL target prediction performance on yeast data. (**A**) Boxplots show the distribution of AUROC values using all genes (blue) and excluding genes on the same chromosome (red) as predictors, for random forest regression (RFR), support vector regression (SVM), ridge regression (RR), and univariate correlation (Corr). (**B**) Scatter plots show AUROC values of classification methods RFR, SVR, and univariate correlation (Corr) vs RR for all variants. Genes on the same chromosome were excluded as predictors for each SNP. (Color figure online)

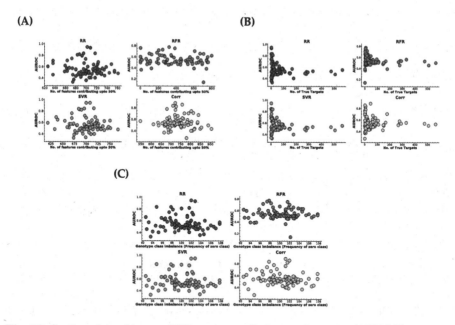

Fig. 10. Scatter plots of trans-eQTL target prediction performance (AUROC) on yeast data against the number of selected model features (**A**), the number of true trans-eQTL targets in the ground-truth network (**B**) and the genotype class balance (frequency of the zero class) (**C**), for random forest regression (RFR), support vector regression (SVR), ridge regression (RR), and univariate correlation (Corr). Genes on the same chromosome were excluded as predictors for each SNP.

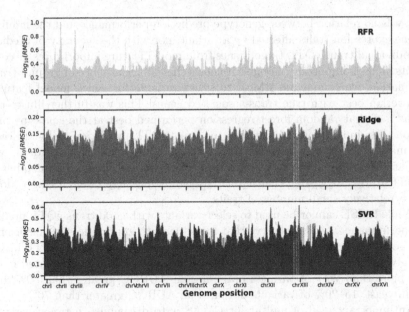

Fig. 11. Expression hotspot maps showing the negative log transformed RMSE values vs genome position for 2884 SNPs in the yeast genome, for random forest (RF, top), ridge regression (Ridge, middle), and support vector regression (SVR, bottom). Genes on the same chromosome were excluded as predictors for each SNP.

again suggesting that non-linear methods like random forest may detect biological signals missed by traditional methods.

4   Discussion

In this study we analyzed the use of machine learning methods for genotype prediction in high-dimensional multi-trait GWAS. The basic hypotheses of reverse genotype prediction from multiple trait combinations are that variants whose genotypes can be predicted with higher accuracy are more likely to have a true effect on a large number of the measured traits, and that feature importances or coefficients in the trained models indicate the strength of association between variants and individual traits. However, existing studies have not presented conclusive evidence for these hypotheses, because they only performed downstream analysis for the highest scoring variants, and only considered linear models. Here we performed an in-depth validation of various machine learning methods for reverse genotype prediction in the context of trans-eQTL analysis, including univariate, ridge regression, random forest, and support vector regression, using both simulated and real transcriptional regulatory networks to define ground-truth sets of trans-eQTL target sets.

Our results support the basic hypotheses only partially. In particular, although genotype prediction performance indeed varied across genetic variants, there was no relation between genotype prediction performance and the number of gene expression traits affected by a variant, nor with the accuracy of predicting individual trans-eQTL target genes from model feature importances or coefficients. This is important, because it shows that in the absence of ground-truth information, we cannot use RMSE to select variants for which model features will overlap best with true trans-associated genes. This was further illustrated by the fact that random forest regression performed best at the genotype prediction task, but performed worst on the trans-eQTL prediction task. The only systematic relation we observed, both in the simulated and the yeast data, was a negative correlation between genotype prediction performance and number of model features, suggesting that if a variant can be predicted well, it can be done with a relatively small number of traits.

While RMSE cannot be used to select variants with good trans-eQTL prediction performance, we did observe that model feature importances or coefficients were generally predictive of how likely a given gene is a true trans-eQTL target of a given variant. Predictive performance was very strong in simulated data, with more than 75% of variants obtaining an AUROC greater than 80%, but also in yeast, 15–20% of variants obtained an AUROC greater than 70%.

An important goal of multi-trait GWAS is to distinguish between variants that are associated with high vs low number of traits. Interestingly, we found that only random forest, but not SVR or ridge regression, resulted in models with a wide variation in the number of selected features across variants. However this involved use of a simple, heuristic strategy for feature selection, and further research to finetune this result will be required.

One aspect of multi-trait GWAS not considered in this study is statistical inference. For linear methods, the null distribution of the model fit score under the assumption of no association can be approximated analytically to obtain a p-value for the significance of any observed score. Non-linear methods such as random forest or SVM require a large number of permutations for each variant separately to obtain a p-value, which becomes computationally infeasible for a large number of variants. However approximate methods may yet overcome this hurdle [16]. More importantly though, since our results indicate that model fit is not related to either the strength or extent of true biological relations, the relevance of performing statistical inference on this test statistic is in doubt.

Another area of future research concerns the generalization to other organisms, in particular human. We focused on realistic simulated data from the DREAM project and data from the eukaryotic model organism yeast, due to the availability of data from a study with extraordinarily large sample size and extensive, high-quality ground-truth transcriptional interaction data. The availability of ground-truth associations also motivated our choice of studying gene expression traits. It will be of interest to expand this work to other types of traits, including protein and metabolite levels, as well as high-dimensional phenotypic traits such as images.

In summary, feature importance weights in machine learning models that predict genotypes from high-dimensional sets of traits identify biologically relevant variant-trait associations, but comparing the relative importance of variants through these models in a GWAS-like manner using a single test statistic remains an open challenge.

References

1. McCarthy, M.I., et al.: Genome-wide association studies for complex traits: consensus, uncertainty and challenges. Nat. Rev. Genet. **9**(5), 356–369 (2008)
2. Manolio, T.A.: Bringing genome-wide association findings into clinical use. Nat. Rev. Genet. **14**(8), 549–558 (2013)
3. Ferreira, M.A., Purcell, S.M.: A multivariate test of association. Bioinformatics **25**(1), 132–133 (2009)
4. Galesloot, T.E., Van Steen, K., Kiemeney, L.A., Janss, L.L., Vermeulen, S.H.: A comparison of multivariate genome-wide association methods. PloS one **9**(4), e95923 (2014)
5. Van Rheenen, W., Peyrot, W.J., Schork, A.J., Lee, S.H., Wray, N.R.: Genetic correlations of polygenic disease traits: from theory to practice. Nat. Rev. Genet. **20**(10), 567–581 (2019)
6. Shabalin, A.A.: Matrix eQTL: ultra fast eQTL analysis via large matrix operations. Bioinformatics **28**(10), 1353–1358 (2012)
7. Qi, J., Asl, H.F., Björkegren, J., Michoel, T.: kruX: matrix-based non-parametric eQTL discovery. BMC Bioinform. **15**(1), 1–7 (2014)
8. Ongen, H., Buil, A., Brown, A.A., Dermitzakis, E.T., Delaneau, O.: Fast and efficient QTL mapper for thousands of molecular phenotypes. Bioinformatics **32**(10), 1479–1485 (2016)
9. O'Reilly, P.F., et al.: MultiPhen: joint model of multiple phenotypes can increase discovery in GWAS. PloS one **7**(5), e34861 (2012)
10. Banerjee, S., et al.: Reverse regression increases power for detecting trans-eQTLs. bioRxiv. (2020)
11. Wang, H., et al.: From phenotype to genotype: an association study of longitudinal phenotypic markers to Alzheimer's disease relevant SNPs. Bioinformatics **28**(18), i619–i625 (2012)
12. Albert, F.W., Bloom, J.S., Siegel, J., Day, L., Kruglyak, L.: Genetics of trans-regulatory variation in gene expression. Elife **7**, e35471 (2018)
13. Monteiro, P.T., et al.: YEASTRACT+: a portal for cross-species comparative genomics of transcription regulation in yeasts. Nucleic Acids Res. **48**(D1), D642–D649 (2020)
14. Pinna, A., Soranzo, N., Hoeschele, I., de la Fuente, A.: Simulating systems genetics data with SysGenSIM. Bioinformatics **27**(17), 2459–2462 (2011)
15. Yates, A.D., et al.: Ensembl 2020. Nucleic Acids Res. **48**(D1), D682–D688 (2020)
16. Knijnenburg, T.A., Wessels, L.F., Reinders, M.J., Shmulevich, I.: Fewer permutations, more accurate P-values. Bioinformatics **25**(12), i161–i168 (2009)

A Non-Negative Matrix Tri-Factorization Based Method for Predicting Antitumor Drug Sensitivity

Carolina Testa$^{(\boxtimes)}$, Sara Pidò , and Pietro Pinoli$^{(\boxtimes)}$

Department of Electronics, Information and Bioengineering, Politecnico di Milano,
Milano, Italy
{carolina.testa,sara.pido,pietro.pinoli}@polimi.it

Abstract. Large annotated cell line collections have been proven to enable the prediction of drug response in the pre-clinical setting. We present an enhancement of Non-Negative Matrix Tri-Factorization method, which allows the integration of different data types for the prediction of missing associations. To test our method we retrieved a dataset from the Cancer Cell Line Encyclopedia (CCLE), containing the connections among cell lines and drugs by means of their IC50 values, and we integrated it by linking cell lines to their respective tissue of origin and genomic profile. We performed two different kind of experiments: a) prediction of missing values in the matrix, b) prediction of the complete drug profile of a new cell line, demonstrating the validity of the method in both scenarios.

Keywords: Non-Negative Matrix Tri-Factorization · Drug sensitivity · Data integration · Drug response prediction

1 Background

Cancer is a highly complex disease due to the enormous level of both intra- and inter-tumor heterogeneity that often displays. Indeed, several tumors of the same organ may vary significantly in important tumor-associated attributes. This is the reason why patients with the same diagnosis can respond in different ways to the same therapy, and this represents the main obstacle to effective treatments [1]. For this reason, it becomes essential to be able to predict if a patient is sensitive or resistant to a specific drug before the administration. Being sensitive to a drug means that the drug manages to have the desired effect on the person, with tolerable side effects; on the contrary, drug resistance represents the inability of the active principle to perform its function. The parameter most extensively used to characterize the response and sensitivity to a drug is the half-maximal inhibitory concentration (IC50), that is the concentration needed to inhibit the 50% of the targeted biological process or component [2]. In particular, in the field of anticancer therapies, the IC50 represents the concentration of drugs needed to kill half of the cells *in vitro*. Since experimental approaches

D. Chicco et al. (Eds.): CIBB 2021, LNBI 13483, pp. 94–104, 2022.
https://doi.org/10.1007/978-3-031-20837-9_8

for the estimation of IC50 values are costly and time-consuming, researchers are increasingly putting efforts into developing computer-based methods for predicting the responsiveness of a patient to a drug. This was made possible thanks to the huge amount of biological, medical and chemical data that have started to be grouped and made publicly available through several tools and databases. In particular, in the context of drug sensitivity, we can certainly cite the Cancer Cell Line Encyclopedia (CCLE) [3] and the Genomics of Drug Sensitivity in Cancer (GDSC) [4] projects, which succeeded in collecting the genetic and pharmacological profile of hundreds of cancer cell lines.

The work of Berrettina et al. [3] can be also considered one of the pioneering machine learning methods proposed for the prediction of sensitive or resistant drug response of a cell line. It exploited CCLE data for a predictive model based on the naïve Bayes classifier. Subsequently, Dong et al. [5] used gene expression features and drug sensitivity data to build SVM-RFE, a wrapper method that firstly performs a feature selection operation and successively uses top features to fit the Support Vector Machine, a supervised learning algorithm for classification. HNMDRP, a network-based method which takes into consideration cell lines, drugs and targets relationships, was then proposed by Zhang et al. [6]. Xu et al. [7] developed the AutoBorutaRF model, which performs a two step feature selection by means of a combination of an autoencoder artificial neural network and the Boruta algorithm, and then uses random forest for classification. More recently, Choi et al. [8] presented a deep neural network model, RefDNN, which pairs molecular structure similarity profiles of drugs and gene expression data of cell lines. In the meanwhile, we can also find DSPLMF, a prediction approach presented by Emdadi et al. [9] based on logistic matrix factorization which allows to compute the probability of cell lines to be sensitive to a drug and thus to classify drug response. To improve the accuracy of the method, gene expression profiles, copy number alterations and single-nucleotide mutation for cell line similarity and chemical structures for drug similarity have been incorporated.

In this scenario fits our work, which has the purpose to address the issue of predicting the sensitivity of a cell line to a drug with a network-based approach based on Non-negative Matrix Tri-Factorization (NMTF), an algorithm designed to factorize an input positive-defined matrix (such as an association matrix of a bipartite graph) in three matrices of non-negative elements. The decomposition has proven to be useful also to predict missing associations. One of the main advantages of NMTF is the possibility to expand the bipartite network integrating several information and thus forming a multi-partite graph; the NMTF algorithm is then used to decompose each of the association matrices, in such a way that the decomposition of each matrix is influenced by the decomposition of the others [10]. The NMTF approach has been used in several domains and in particular it demonstrated to have elevated performances in both finding new indications for approved drugs and new synergistic drug pairs, in particular when including several heterogeneous data types [10,11]. The main focus of this work is to adapt the model to predict the sensitivity of a cancer cell line to a set of anti-tumor drugs integrating the associations between cell lines and drugs with tissue and gene expression-related data. In the context of precision medicine, the prediction of drug response and sensitivity based on genetic features is becoming of fundamental relevance to speed the emergence of 'personalized' therapeutic regimens. Being able to determine a priori to which drugs a patient, with its genomic features, is sensitive or resistant would save a lot of precious time and improve the efficiency of the therapy.

2   Material and Methods

2.1   Datasets

For our experiments, we used the dataset retrieved from the Cancer Cell Line Encyclopedia (CCLE) [3] which comprised the association among 1065 cell lines and 266 antitumor drugs, measured in terms of IC50. In light of the presence of a large amount of missing values in the dataset, we firstly performed a filtering operation that allowed us to reduce them from 20% to 2%, by eliminating both cell lines and drugs with more than the 50% of missing data. Subsequently, we binarized the matrix: since the IC50 is representative of the response of a cell line to a drug, we considered a cell line to be sensitive to a drug if the corresponding IC50 value was lower than a threshold and, on the contrary, a cell line was considered resistant if that value was higher. As threshold for classification we selected the median of IC50 values of each drug, considering all the cell lines. From CCLE we have been able to retrieve also further datasets containing additional information; in particular, we took into consideration tissues of origin of the tumors and the gene expression profiles quantified by RNA-seq experiments. After processing and integrating all these data, as described in detail in Sect. 2.2, we obtained a final dataset containing 379 cell lines and 202 drugs.

2.2   Model

In order to integrate all the available information, we modeled the set of cell lines C, the set of drugs D, the set of tissues T and the set of genes G as the multipartite network in Fig. 1, where each cell line is connected to the drugs to which it is sensitive, the tissue of origin and a set of genes, with the weight of the edge representing the expression of the gene in the cell line.

Such network is equivalent to the set of its association matrices: a binary matrix X_{CD} connecting cell lines to drugs, a binary matrix X_{TC} connecting cell lines to tissues, and a real matrix X_{GC} connecting cell lines to genes. We built the three matrices as follows:

- we represented the $IC50$ data as a matrix $X \in \overline{\mathbb{R}}_{\geq 0}^{|C| \times |D|}$, being $\overline{\mathbb{R}}_{\geq 0} = \mathbb{R}_{\geq 0} \cup \{+\infty\}$, such that $X[i,j]$ indicates the IC50 value of the j-th drug on the i-th cell line if a measure is available, or $+\infty$ otherwise. We transform X into the binary matrix $X_{CD} \in \{0,1\}^{|C| \times |D|}$, such that:

$$X_{CD}[i,j] = \begin{cases} 1 & \text{if } X[i,j] < M_j \\ 0 & \text{otherwise} \end{cases}$$

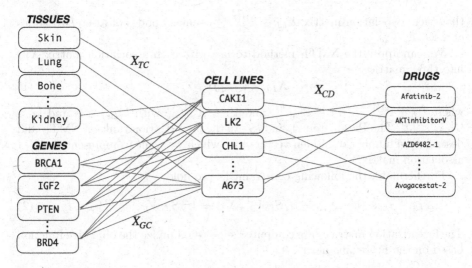

Fig. 1. Multipartite graph connecting Tissues, Gene expression, Cell lines and Drugs. The three association matrices of the graph are also indicated.

where M_j is the median of the IC50 values for the j-th drug that are different from $+\infty$. This step is necessary as different drugs work on different scale of dosage. In other words, we define the i-th cell line to be sensitive to the j-th drug if the IC50 value $X[i, j]$ is lower than the median for that particular drug;

– we built the matrix $X_{TC} \in \{0, 1\}^{|T| \times |C|}$ that connects cell lines to tissues as:

$$X_{TC}[c, t] = \begin{cases} 1 & \text{if } c \text{ belongs to the tissue } t \\ 0 & \text{otherwise} \end{cases}$$

– we considered the matrix $X'_{GC} \in \mathbb{R}^{|G| \times |C|}$, where G is a set of genes and $X'_{GC}[g, c]$ represents the $RPKM$, that means *reads per kilobase of transcript, per million mapped reads*, measured for the given gene g in the specific cell line c, by a RNA-seq experiment. To select a valuable subset of genes, we retrieved the 1,000 genes with the highest standard deviation of the expression across the cell lines. Then, for each drug (that is each column in the matrix X_{CD}) we performed a LASSO [12] feature selection. Finally, we kept into consideration only the 532 genes that are selected as predictive features for at least two drugs and build the matrix $X''_{GC} \in \mathbb{R}^{532 \times |C|}$. Finally we performed a minmax scaling on the columns of the matrix and considered the $X_{GC} \in [0, 1]^{532 \times |C|}$ matrix.

2.3   Method

Let's consider a multipartite graph $\mathcal{G}$; for the purpose of this work, we can represent the graph as a set of association matrices, that is $\mathcal{G} = \{X_{IJ}\}$, such that each association matrix $X_{IJ} \in \mathbb{R}_{\geq 0}^{|I| \times |J|}$ connects nodes of a set I to nodes of a set J.

We can apply the NMTF method to factorize each association matrix X_{IJ} into three matrices:

$$X_{IJ} \cong U_I S_{IJ} V_J^\top \tag{1}$$

where $U_I \in \mathbb{R}_{\geq 0}^{|I| \times k_i}, S_{IJ} \in \mathbb{R}_{\geq 0}^{k_i \times k_j}$, and $V_J \in \mathbb{R}_{\geq 0}^{|J| \times k_j}$ with $k_i, k_j \in \mathbb{N}$ and $k_i < |I|, k_j < |J|$. The Parameters k_i and k_j are the factorization ranks of NMTF and describe the number of hidden vectors into which we want to represent the X_{IJ} association matrix.

Furthermore, the following constraint has to hold:

$$\forall X_{IJ}, X_{JL} \in \mathcal{G}, \quad X_{IJ} \cong U_I S_{IJ} V_J^\top, X_{JL} \cong U_J S_{JL} V_L^\top \implies V_J \equiv U_J \tag{2}$$

The factorization matrices are computed so as to minimize the objective function based on the Frobenius norm:

$$\mathcal{L}(\mathcal{G}|\Theta) = \sum_{X_{ij} \in \mathcal{G}} \left\| X_{ij} - U_i S_{ij} V_j^\top \right\|_{Fro}^2 \tag{3}$$

where Θ represents the set of all the factorization matrices.

A minimum of the objective function can be computed algorithmically by (a) initializing the factorization matrices and (b) applying the following multiplicative update rules:

$$U_I \leftarrow U_I \odot \frac{\sum_Q X_{IQ} V_Q S_{IQ}^\top + \sum_Q X_{QI}^\top U_Q S_{QI}}{\sum_Q U_I S_{IQ} V_Q^\top V_Q S_{IQ}^\top + \sum_Q U_I S_{QI}^\top U_Q^\top U_Q S_{QI}} \tag{4}$$

$$V_J \leftarrow V_J \odot \frac{\sum_Q X_{QJ}^\top U_Q S_{QJ} + \sum_Q X_{JQ} V_Q S_{JQ}^\top}{\sum_Q V_J S_{QJ}^\top U_Q^\top U_Q S_{QJ} + \sum_Q V_J S_{JQ} V_Q^\top V_Q S_{JQ}^\top} \tag{5}$$

$$S_{IJ} \leftarrow S_{IJ} \odot \frac{U_I^\top X_{IJ} V_J}{U_I^\top U_I S_{IJ} V_J^\top V_J} \tag{6}$$

where $\odot$ and $\frac{\bullet}{\bullet}$ stand for Hadamard element-wise multiplication and division, respectively. Updating rules must be iteratively calculated. We perform 100 warm-up iteration and then we iterate until a stop criterion is met; in our experiments we used $\frac{\mathcal{L}^{i-1} - \mathcal{L}^i}{\mathcal{L}^{i-1}} < 10^{-6}$, where $\mathcal{L}^{i-1}$ and $\mathcal{L}^i$ are respectively the values of the loss function after the last and the previous iterations [13].

For matrices initialization, which is a critical aspect of the method, we adopted a k-means approach [14–16].

2.4   Prediction of Novel Associations

The prediction of novel associations between two sets of nodes can be interpreted as a matrix completion task. The NMTF method is applied in order to predict novel links between two classes of nodes. In particular, we focused on the associations between cell lines and drugs. After that

$$\tilde{X}_{CD} = U_C S_{CD} V_D^\top$$

has been computed with the following updating rules

$$U_C \leftarrow U_C \odot \frac{X_{CD} V_D S_{CD}^\top}{U_C S_{CD} V_D^\top V_D S_{CD}^\top} \tag{7}$$

$$V_D \leftarrow V_D \odot \frac{X_{CD}^\top U_C S_{CD}}{V_D S_{CD}^\top U_C^\top U_C S_{CD}} \tag{8}$$

$$S_{CD} \leftarrow S_{CD} \odot \frac{U_C^\top X_{CD} V_D}{U_C^\top U_C S_{CD} V_D^\top V_D} \tag{9}$$

we applied a threshold τ, typically $0 < \tau < 1$, and we considered that the i-*th* cell line is associated with the j-*th* drug if the predicted value $\tilde{X}_{CD}[i,j] > \tau$.

2.5   Prediction of the Whole Drug Profile for a New Cell Line

Another scenario is when a novel cell line is included in the network. In this situation, while we know the genetic feature of the cell line and its tissue of origin, we do not have information about the drugs to which it is sensitive.

We here propose a slight modification of the NMTF multiplicative update rules, in order to being able to predict the complete drug profile for the novel cell line. Since we have no correct information in the matrix we aim to reconstruct for the novel cell line, we do not consider the influence of X_{CD} during the update of U_C matrix. Thus, the new rules to update U (for our network) were:

$$U_C \leftarrow U \odot \frac{X_{TC}^\top U_T S_{TC}}{U_C S_{TC}^\top U_T^\top U_T S_{TC}} \tag{10}$$

$$U_C \leftarrow U \odot \frac{X_{GC}^\top U_G S_{GC}}{U_C S_{GC}^\top U_G^\top U_G S_{GC}} \tag{11}$$

when only X_{TC} or X_{GC} are taken into account, while

$$U_C \leftarrow U \odot \frac{X_{TC}^\top U_T S_{TC} + X_{GC}^\top U_G S_{GC}}{U_C S_{TC}^\top U_T^\top U_T S_{TC} + U_C S_{GC}^\top U_G^\top U_G S_{GC}} \tag{12}$$

when both X_{TC} and X_{GC} are added to the network. Updating rules for V and S remained unvaried with respect to 8 and 9.

3   Results

Here, we report the results of different trials that we performed on the dataset illustrated in Sect. 2.1. In particular, we apply NMTF method, illustrated in Sect. 2.3, for two different tasks: the prediction of novel cell line-drug associations and the prediction of the drug profile for a new cell line. We evaluate our results using the AUROC (that means *area under the receiver operating characteristic curve*) and the comparison between the actual IC50 values of pairs predicted sensitive ($\tilde{X}_{CD} > 0.6$) and predicted resistant ($\tilde{X}_{CD} < 0.4$).

3.1   Prediction of Novel Associations

In order to validate the model, we apply a mask that covers randomly the 5% of the association matrix X_{CD}. We run the method on the single matrix X_{CD} without passing other information and we compute the evaluation metrics, testing various combination of the parameter $k1$ and $k2$. The best configuration corresponds to $k1 = 25$ and $k2 = 15$. With these parameters the model performs well and leads to a AUROC equal to 0.84417 as shown in Fig. 2a. On the best configuration we run also the Welch test, a two-sample location test which is used to test the hypothesis that two populations have equal means and is more reliable when the two samples have unequal variances and possibly unequal sample sizes. Comparing the two box plots, the predicted sensitive and resistant associations appear to be significantly different (p-val $\approx$ 0.0). Results are shown in Fig. 2b.

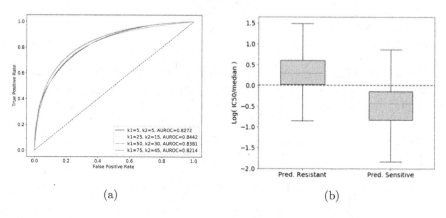

(a)   (b)

Fig. 2. Performances using a random mask on X_{CD}; (a) reports the ROC curve with several values of k1 and k2, where k1 and k2 are the factorization ranks of the NMTF. The best configuration corresponds to k1 = 25 and k2 = 15; (b) Boxplots of the IC50 values, divided by the means, of the predicted sensitive and resistant pairs using a random mask on X_{CD}, with the best configuration.

3.2   Prediction of the Whole Drug Profile for a New Cell Line

In this case, we apply a mask on a single row of the matrix X_{CD} in order to simulate the addition of a novel cell line.

Considering *only* X_{CD} matrix does not provide meaningful results, as shown in Fig. 3a, 3b. As expected, without any additional information, the AUROC is 0.50506, and the two classes are not different. This result proves that it is impossible to predict a complete drug profile for a novel cell line without considering other data.

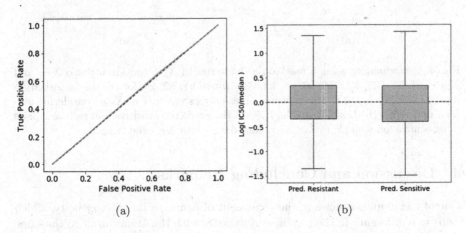

(a)   (b)

Fig. 3. Performances using a mask on a single row of X_{CD} with k1 = 25 and k2 = 15, where k1 and k2 are the factorization ranks of NMTF. (a) ROC curve using a mask on a single row on X_{CD}. (b) Boxplots of the IC50 of the predicted sensitive and resistant pairs using a mask on a single row of X_{CD}.

Thus, we tested the method by also adding the X_{TC} matrix alone, X_{GC} matrix alone as well as the two together.

The AUROCs in Fig. 4a proves that adding information increases the performances of the predictor. Including the tissue of origin, the method is able to reach an AUROC = 0.60244. If also gene expressions are added to the model, we observe a significant improvement (AUROC = 0.71063). Finally, when both gene expressions and tissues of origin are considered, and the AUROC increases to 0.71163. In Fig. 4b the comparison between predicted resistant and sensitive drugs, when all the information is used, is shown; the Welch test confirms the difference in the distribution of the two classes (p-val = 9.30232×10^{-18}), with the IC50 of the predicted sensitive drugs clearly below the median.

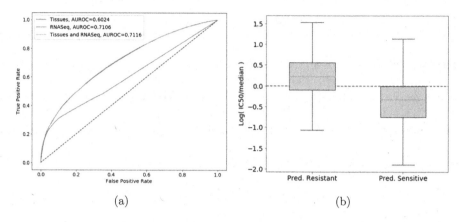

(a)   (b)

Fig. 4. Performances using a mask on a single row of X_{CD} considering also X_{TC} and X_{GC} with k1 = 25, k2 = 15, k3 = 5, k4 = 30, where k1, k2, k3, k4 are the factorization ranks of NMTF. (a) ROC curves using a mask on a single row of X_{CD} considering also X_{TC} and X_{GC}. (b) Boxplots of the IC50 of the predicted sensitive and resistant pairs using a mask on a single row of X_{CD} considering also X_{TC} and X_{GC}.

4   Discussion and Concluding Remarks

One of the main obstacles in the treatment of cancer is its heterogeneity, which leads to a difference in the response of patients with the same cancer to the same drug [17,18]. In this context, computer-based approaches can be very powerful tools in order to identify in advance which drugs a patient is sensitive to and to which drugs does not respond instead [19]. To reach this goal, we proposed a network-based method which exploits Non-Negative Matrix Tri-Factorization algorithm for the prediction of the sensitiveness of a patient, which is represented by the cell line extracted from his tumor mass, to a drug. We performed the experiments on a dataset retrieved from CCLE, which contains cell lines and antitumor drugs linked by means of their IC50 values. In our work, we demonstrated that predicting the sensitivity of a specific drug for a given cell line for which many IC50 experiments are available is a rather easy task. In our experiments, using plain NMTF method without additional information for this task allows to reach high performances (AUROC = 0.84417). On the contrary, predicting drug sensitivity profile for a novel cell line is more complex: indeed, NMTF method without other data scores as bad as a random predictor.

To overcome this limitation, we proposed a two-fold solution: (a) we developed an improved version of NMTF algorithm, which generates predictions taking into account only meaningful information, and (b) we integrated other information, namely the tissues of origin and the gene expressions of the corpus of cell lines. When all the available data are provided, the proposed method shows much better performances: the resulted AUROC is equal to 0.71163.

Our results suggest that NMTF is a valid method for the prediction of sensitiveness and resistance of a patient to a drug. In particular, the method gives very high results for the matrix completion task, meaning that with this approach is easy to predict novel sensitivity or resistance associations to missing drugs, even without adding further information to the primary association matrix. Instead, for new cell lines with no previous connections to drugs the prediction is a little more complicated and the link with more data matrices is needed. Indeed, the initial value of prediction is quite low, but it increases adding patient related data. In particular, it has only a 5% increase adding just the tissues matrix, while the addition of gene expression data leads to a higher 10% increase of the AUROC value. However, the employment of both matrices causes a slight increase of the AUROC value with respect to the use of gene expression data alone. This confirms the hypothesis that, since each cell line is linked only to one tissue, information about tissues are poorly informative and supply a minor contribution to the prediction compared to gene expression data.

Finally, to test the effect and the need of NMTF with respect to a baseline method, we computed a leave-one-out validation for 202 binary logistic regressors (one for each drug). Each predictor uses as feature the gene expression of a cell and the one-hot-encoding of the tissues, and as label the response of the cell for the drug associated to the regressor. The average AUROC of this experiment is 0.69128, thus performing almost the 3% worse than NMTF.

As future development we would like to enlarge the network to further improve the performance. Moreover, we want to implement a regression method in order to being able to predict also the weight of the connection, that means the IC50 value.

To conclude, we believe that our method could certainly help to find more rapidly the right therapy for the patient, saving time and providing the best treatment from the start, which is one of the most critical part in the discovery of the correct therapeutic plan of a person. Indeed, for a patient with cancer, time is the most important resource and a "trial-and-error" approach is not the most advantageous way to proceed in finding the right cure. A priori knowledge of which drug will work and which will not on each specific patient should become one of the fundamental strongholds in the context of precision medicine based treatments.

Acknowledgments. Supported by the ERC Advanced Grant 693174 "Data-Driven Genomic Computing" (GeCo).

References

1. De Lartigue, J.: Tumor heterogeneity: a central foe in the war on cancer. J. Commun. Supp. Oncol. **16**(13), E167–E174 (2018)
2. Neubig, R.R., Spedding, M., Kenakin, T., Christopoulos, A.: International union of pharmacology committee on receptor nomenclature and drug classification. xxxviii. update on terms and symbols in quantitative pharmacology. Pharmacol. Rev. **55**(4), 597–606 (2003)
3. Barretina, J., et al.: The cancer cell line encyclopedia enables predictive modelling of anticancer drug sensitivity. Nature **483**(7391), 603–607 (2012)

4. Yang, W., et al.: Genomics of drug sensitivity in cancer (GDSC): a resource for therapeutic biomarker discovery in cancer cells. Nucleic Acids Res. **41**(D1), D955–D961 (2012)

5. Dong, Z., et al.: Anticancer drug sensitivity prediction in cell lines from baseline gene expression through recursive feature selection. BMC Cancer **15**(1), 1–12 (2015)

6. Zhang, F., Wang, M., Xi, J., Yang, J., Li, A.: A novel heterogeneous network-based method for drug response prediction in cancer cell lines. Sci. Rep. **8**(1), 1–9 (2018)

7. Xiaolu, X., Hong, G., Wang, Y., Wang, J., Qin, P.: Autoencoder based feature selection method for classification of anticancer drug response. Front. Genet. **10**, 233 (2019)

8. Choi, J., Park, S., Ahn, J.: RefDNN: a reference drug based neural network for more accurate prediction of anticancer drug resistance. Sci. Rep. **10**(1), 1–11 (2020)

9. Emdadi, A., Eslahchi, C.: DSPLMF: a method for cancer drug sensitivity prediction using a novel regularization approach in logistic matrix factorization. Front. Genet. **11**, 75 (2020)

10. Ceddia, G., Pinoli, P., Ceri, S., Masseroli, M.: Matrix factorization-based technique for drug repurposing predictions. IEEE J. Biomed. Health Inf. **24**(11), 3162–3172 (2020)

11. Pinoli, P., Ceddia, G., Ceri, S., Masseroli, M.: Predicting drug synergism by means of non-negative matrix tri-factorization. IEEE/ACM Trans. Comput. Biol. Bioinf. **19**, 1956–1967 (2021)

12. Tibshirani, R.: Regression shrinkage and selection via the lasso. J. Roy. Stat. Soc.: Ser. B (Methodol.) **58**(1), 267–288 (1996)

13. Čopar, A., Zupan, B., Zitnik, M.: Fast optimization of non-negative matrix tri-factorization. PloS One **14**(6), e0217994 (2019)

14. Ding, C., Li, T., Peng, W., Park, H.: Orthogonal nonnegative matrix t-factorizations for clustering. In: Proceedings of the 12th ACM SIGKDD International Conference on Knowledge Discovery and Data Mining, pp. 126–135 (2006)

15. Wild, S., Curry, J., Dougherty, A.: Improving non-negative matrix factorizations through structured initialization. Pattern Recogn. **37**(11), 2217–2232 (2004)

16. Xue, Y., Tong, C.S., Chen, Y., Chen, W.S.: Clustering-based initialization for non-negative matrix factorization. Appl. Math. Comput. **205**(2), 525–536 (2008)

17. Marusyk, A., Polyak, K.: Tumor heterogeneity: causes and consequences. Biochimica et Biophysica Acta (BBA)-Rev. Cancer **1805**(1), 105–117 (2010)

18. Melo, F.D.S.E., Vermeulen, L., Fessler, E., Medema, J.P.: Cancer heterogeneity–a multifaceted view. EMBO Rep. **14**(8), 686–695 (2013)

19. Chen, Y., Juan, L., Lv, X., Shi, L.: Bioinformatics research on drug sensitivity prediction. Front. Pharmacol. **12**, 799712 (2021)

A Rule-Based Approach for Generating Synthetic Biological Pathways

Joshua Thompson[1], Haoyu Dong[2], Kai Liu[2], Fei He[2(✉)], Mihail Popescu[1], and Dong Xu[1]

[1] University of Missouri, Columbia, MO 65201, USA
jltmh3@mail.missouri.edu, PopescuM@health.missouri.edu,
xudong@missouri.edu
[2] Northeast Normal University, Changchun 130024, China
{donghy300,liuk479,hef740}@nenu.edu.cn

Abstract. Deep learning has recently enabled many advances for computer vision applications in image recognition, localization, segmentation, and understanding. However, applying deep learning models to a wider variety of domains is often limited by available labeled data. To address this problem, conventional approaches supplement more samples by augmenting existing datasets. However, these up-sampling methods usually only create derivations of the source images. To supplement with unique examples, we introduce an approach for generating purely synthetic data for object detection on biological pathway diagrams, which describe a series of molecular interactions leading to a certain biological function based on a set of rules and domain knowledge. Our method iteratively generates each pathway relationship uniquely. These realistic replicas improve the generalization significantly across a variety of settings. The code is available at https://github.com/JRunner97/Pathway_Data_Synthesis.

Keywords: Synthetic data · Data augmentation · Biological pathways · Object detection

1 Scientific Background

1.1 Introduction

Much of the progress in object detection from the past decade can be attributed to improved architecture design (RCNN [1], Fast-RCNN [2], Faster-RCNN [3], RetinaNet [4], etc.) and increased access to more high-quality data. For example, the COCO [5] 2017 dataset alone contains over 100k labeled training images across 80 classes. However, in some domains, especially the biomedical field, high-quality labeled data is often expensive and time-consuming to collect. To address this issue, many up-sampling methods have been proposed to increase dataset sizes. Common approaches to this issue involve creating more data through simple augmentations. However, recently there has been more interest in generating entire synthetic samples. Our work is in this line of research. We propose a general method for iteratively up-sampling diagram/figure data.

© The Author(s), under exclusive license to Springer Nature Switzerland AG 2022
D. Chicco et al. (Eds.): CIBB 2021, LNBI 13483, pp. 105–116, 2022.
https://doi.org/10.1007/978-3-031-20837-9_9

As an exemplary task, we target localizing gene-gene interactions in biological pathway figures. The ability to automatically extract such relationships can be very useful in summarizing diagram/paper contents and by extension speed up article curation efforts. To that end, the genes and their relationships need to be localized to convert gene-gene relations into a computational format. Nowadays, deep learning-based models are dominant in image object localization (or named object detection). Yet, the mainstream deep object detection models are labeled data hungry, which requires annotating the bounding boxes of objects from the images. Such annotation is costly and labor intensive. In this work, we show how to easily generate each gene-gene interaction uniquely with annotation to produce realistic pathway diagrams to supplement a small, annotated set for training a deep learning model to recognize genes and their relationships from pathway figures. With such a deep model, a tool reconstructing the graphs of pathways will be further explored.

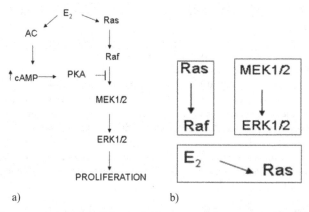

Fig. 1. a. Shows an example pathway figure. b. shows several gene-gene relationships with their bounding boxes we could sample from this image. Text entities in these figures are genes or other biological components. Arrows represent gene activations and t-bars indicate gene inhibitions.

1.2   Related Work

When needing to up-sample a dataset, the first step for most is to use traditional data augmentations. Such methods usually focus on positional modifications such as random flips, scaling, cropping, rotations, and translations. This kind of augmentation helps make models capture similar signals but from different viewpoints. Other types of augmentation focus on changing color characteristics such as lighting, contrast, hue, and saturation. This kind of augmentation is helpful to make models more color agnostic and focus on shape features. Involved modifications, such as kernel filters, random erasing, and injecting random noise, can also be beneficial. These methods can promote a model to use more contextual information and leverage larger-scale features. An overview of these methods is provided in [6]. The key limitation of such augmentations is that they do not significantly modify the underlying training signal, but instead create derivations of the same one. This can prevent generalization to similar data but in different styles, orientations, or settings. Another approach to up-sampling does not just modify existing images but instead creates entirely new ones. This approach often leverages deep learning methods such as conditional GANs [7], adversarial training, and neural style transfer. Synthetic data generation has previously been applied in generating 3D point clouds for training [8, 9]. Our work is complementary to other synthetic data generation methods and targets the object detection task specifically based on a set of rules. Our rule-based method does not require significant training data preparation and deep learning background; hence, it is more convenient and biology-aware than related deep learning approaches.

2   Materials and Methods

2.1   Synthetic Data Generation

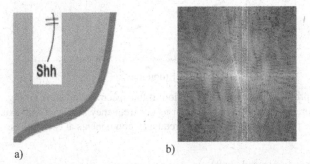

a)   b)

Fig. 2. An example showing the candidate region and its radial profile for filtering high-frequency components. (a) a sample slice of a pathway image; (b) radial profile of the slice in the spectral domain by applying the fast-Fourier transformation.

While simple augmentations can increase the total number of training images several times over, they do not increase the number of unique relationships. It is observed that this may lead to diminishing returns in generalization as the same relationships are repeatedly seen during training. Our approach generates fully synthetic samples. Since many relationships in pathway diagrams follow a simple structure, they are easy to reproduce. A few components make up a relationship in a pathway figure: two entities and a connecting identifier. These identifiers can be represented as arrows, objects, or proximity. Here we will first target two indicators frequently encountered in our pathway figures: arrows (activate) and t-bars (inhibit), as shown in Fig. 1. We generate these fully synthetic relationships by first identifying an empty region on a template image. As we generate new images, we cannot place the slices randomly on the templates, as there is a significant potential for overlap that does not exist in the ground truth figures. We first sample a random candidate region on a template (Fig. 2.a) where our relationship could be placed. Then, we convert that destination region to the spectral domain (Fig. 2.b) via a fast-Fourier transform [13] and calculate the radial profile of that slice in the spectral domain (Fig. 3). The radial profile is the sum over pixel values the same radius away from the slice's center. We do this to see how many high-frequency components exist in that destination region since the increased radius corresponds to higher frequencies in the source image. In images, high-frequency components correspond to edges, contrast, and complex shapes. Our intuition is that pre-placed slices on the image will then be represented in the high-frequency regions. We can effectively search for good placements on the templates by setting a threshold on those high-frequency components.

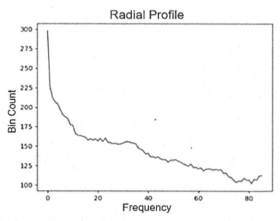

Fig. 3. Radial profile for the candidate region in the spectral domain from Fig. 2b. The x-axis shows the pixel radius from the radial center as our frequency analog. The y-axis represents the number of white pixels that points to high-frequency components at each radius.

With a region selected, we then determine our entity placement given the area's dimensions (see Fig. 4.a). Next, we draw a spline between the two textboxes (Fig. 4.b) and add an indicator head at one of the spline ends (Fig. 4.c). This class identifier can be an arrow or a t-bar (for activating or inhibiting relationships). This approach effectively mimics the structure of many pathway relationships. Using such fully synthetic data generation has several advantages over traditional augmentation for this task. The model can train on more diverse training data since no repeated relationships exist. This enables us to target specific types of exotic relationships that are more difficult to categorize due to little data: (e.g., curvy arrows, splines with corners, or dashed splines). Additionally, this process can be multi-threaded to generate many diagrams at once.

2.2 Implementation Details

Checking Background. While simple to outline, implementing each step is more involved. In the case that all pixels in the destination region are the same, we do not have to run the full spectral check and can immediately stitch a relationship. When this is not the case, we convert the destination region to the spectral domain and generate its radial profile, as previously mentioned. Notably, when calculating the radial profile on this output, we must normalize by distance to the center of the region since as the radius grows so does the number of pixels at that radius. To filter out regions with too

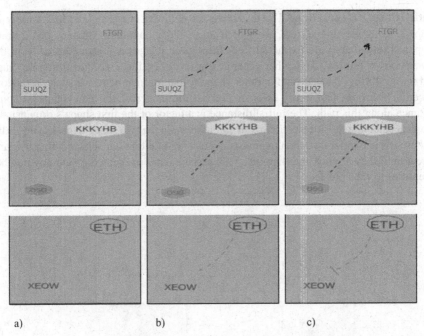

a)   b)   c)

Fig. 4. To generate a relationship, with two placed entities (a), we denote their relationship by drawing a spline/dashed line between them (b) and placing an identifying indicator at one end (c).

many high-frequency components, we look at the binned statistics over the radial profile. Specifically, we bin the radial profile into 4 sections and look at the 2 later bins corresponding to the higher frequencies. If the binned mean for either of those regions is too large, we can rule out this placement. We used a threshold of 50 to filter out slices with too many high-frequency components to determine our placement. This specific threshold balances allowing color gradients while still removing any slices with harsh edges. We found this method to be more effective than simply looking at the pixel statistics of the destination slice and setting a threshold for the standard deviation, while other approaches failed on edges of similar pixel values. Using our method produced more realistic figures that better resembled the source dataset.

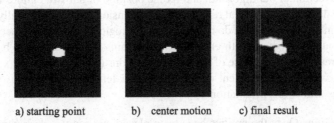

a) starting point   b)   center motion   c) final result

Fig. 5. To generate a cluster of entities, we start from the shape masks of two entities. We iteratively move one shape's center until the IoU between the shapes is 0.

Entity/Cluster Generation. The text boxes of pathway entities come in a variety of shapes. To mimic this variety, we pull from a folder containing images of arbitrary shapes. These shapes are extracted from the source images and transformed to fit the dimensions of our text. To further generate clusters of entities, we start from one shape and its mask (Fig. 5.a). To add another shape to the arrangement, we select a random direction from the first entity's center and set the center of the new shape as some distance along this path. The initial distance is a factor of the first shape's dimensions. We then calculate the intersection-over-union between the two shape's masks (Fig. 5.b). If there is any overlap, we then increase the push factor along our selected direction and repeat until they are non-overlapping. This process can be repeated to add any number of entities to each cluster.

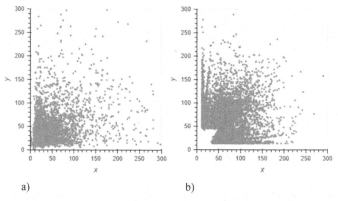

a)   b)

Fig. 6. Scatter plots showing the distribution of relationship sizes in real pathways 6.a and our synthetic pathways 6.b. The x-axes and y- axes represent the width and height of relationship bounding boxes in the pathways, respectively.

Entity Placement. To effectively replicate the entity positions seen in real pathway diagrams, we look at the histogram of relationship dimensions seen in those figures. Figure 6.a shows a 2D-histogram of 1000 diagrams, highlighting that those relationships are most concentrated in the low dimensional regimes. We use this distribution to guide the region selection of our algorithm and fix the position of our two entities to opposite corners. We do this to let the real relationship's dimensional distribution fully guide our entity placement. However, we do maintain that the region selected for placement must be large enough to contain both entities. This explains the gap in Fig. 6.b that is not seen in the real data. In the case of extreme dimensions (e.g., much larger x than y and vice versa), we fix the placement to be top-down or left-right. We do this since most relationships follow this orientation.

Drawing Spline. Once the entities have been placed, we can use their centers as reference points for the connecting spline. We use three different types of splines: lines, arches, and corners. For direct lines, we must first find the start and end points for the spline by interpolating a direct line between the centers of the two text boxes. We then select the n-th point along the line outside the textboxes as the respective start and end points. We set n dynamically based on the distance between the two text boxes. With the start and end points, we then draw a line between these two anchor points as our spline. If we draw an arch, we start with the same method for obtaining the start and end points. Then, we calculate the slope perpendicular to the line between them. With this, we can calculate a third anchor point that will mark the arch's apex. Using all three anchor points, we can then interpolate a spline between them. We set a parameter to determine how far from the baseline the third reference point should be. This allows us to control how 'curvy' each arch is. When drawing a cornered spline, we use a different method for obtaining the start and end points. For each square configuration, we can have two different placements for the start and end points. They can be placed outside the textboxes at some pre-set distance towards the same empty corner. Then we use their max or min dimensions to determine a corner point for the third anchor and connect these three points. To draw dashed splines instead, we simply omit placing intervals of the spline. These intervals are drawn from random bounds and depend on the spline's thickness.

Drawing Indicator. With the spline drawn, we now place an indicator head at one of the ends. Since there are different styles of indicators as well, we follow the same approach as entity shapes and pull from a set of indicator shape images. Again, we extract and transform the indicators as needed for the given spline style. We also rotate these indicators to follow the slope of the spline near the end point and place that transformed indicator onto the end of the spline.

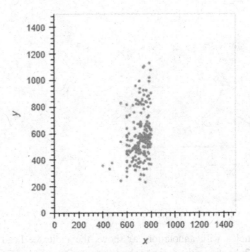

Fig. 7. Histogram Scatter plot showing the distribution of pathway image dimensions. The x-axes and y- axes represent the width and height of relationship bounding boxes in the pathways, respectively.

Parameter Configuration. The above methods detail how to generate a new relationship, but we can control many parameters for this process to ensure that each one is sufficiently unique. For each label, we can control the font color, style, size, and thickness. For each textbox, we can control the textbox margin, background color, textbox shape, and border thickness. For each spline, we can control the indicator placement, type, length, width, color, and thickness. We dynamically change these parameters during generation. This enables us to generate a more robust dataset that mimics the wide variety of relationships seen in real data. For instance, we use the image dimension distribution from real pathways (Fig. 7) to set the bounds for our templates.

3   Experimental Setup

3.1   Model

To validate the contribution of the proposed method to gene-gene relationship localization modeling, we evaluated our method on the widely used RetinaNet [4] architecture with a ResNet-50 [11] backbone pre-trained on the ImageNet [12] dataset. In all of our experiments, we fine-tuned this model for 50 epochs with a learning rate of 0.01. During training RetinaNet with our synthesized images, we are limited to 1 image per batch due to its minimal 8 gigabytes GPU memory requirement. For loss, we used a combination of the sigmoid focal loss [4] for classification (for imbalanced class distributions) and dense box regression for localization.

3.2   Data

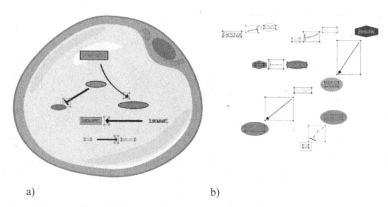

a)   b)

Fig. 8. Synthetic samples with annotation. a) shows the indicator head annotations used in Experiments 4.1 & 4.2. b) shows the indicator body annotations used in Experiment 4.3.

The base augmented dataset was generated from 250 annotated pathway diagrams collected from KEGG [10]. These base figures include 3326 activate indicators, 637 inhibit indicators, and 6461 textboxes. We used salt & pepper, color correction, and random noise to generate 4,000 new training images. We treated training with augmented data alone as our benchmark. For the synthetic data, we generated three sets of increasing size with 1,000, 3,000, and 6,000 images (Fig. 8 as an example). We then looked at how different amounts of synthetic data coupled with the augmented dataset can improve the generalization of our model. We evaluated 45 images held-out from the ground truth data as our validation set and examined the mean-average-precision (mAP) over the three classes: inhibit indicators, activate indicators, and gene text. The mAP measurement captures how well all objects are detected and classified. Results displayed in the tables are averages and standard deviations over three runs to demonstrate the stability of the proposed method.

4 Results

4.1 Synthetic Data for Mixed-Batches

Table 1. Validation mAP for increasing amounts of synthetic data used.

Training data	Aug	Aug + 1k Syn	Aug + 3k Syn	Aug + 6k Syn
mAP	26.9 $\pm$ 2.9	26.5 $\pm$ 0.4	28.8 $\pm$ 2.4	**29.1 $\pm$ 2.1**

In this set of experiments, we look at how our method can affect the generalization of the model (shown in Table 1). With the 4000 augmented samples as our baseline, we see how increasing the amount of synthetic data used in mixed batches affects validation. We find that our method used in conjunction with the augmented data improves the generalization of RetinaNet. This is likely because we can introduce unique relationships/features that simple augmentations cannot. This notion is supported by the fact that increasing the amount of synthetic data generated by our method continues to improve the performance of the model.

4.2 When to Use Synthetic Data

Following the improvement shown by leveraging synthetic data in mixed batches, we sought to understand how different combinations of real and synthetic data affect training. To that end, we first measured the performance of the augmented data and different amounts of synthetic data on the independent set of 45 images using mAP: 50 as shown in Fig. 9. Increasing the amount of synthetic data, as expected, showed improvement in generalization from 3k at 30.2 to 20k at 32.4 mAP. However, both were unable to fully close the gap to the augmented data 41.3 mAP. Interestingly, if we first pretrain on the augmented data with a learning rate of 0.01 and finetune with synthetic data at 0.005, we also see improvement over augmented training alone at 44.6 mAP. This improvement is even more pronounced if we include the augmented data in our finetune stage and use mixed batches (reaching 49.3). We flipped this experiment to pretrain with synthetic and finetune with mixed batches and can see another boost in performance (50.7 mAP

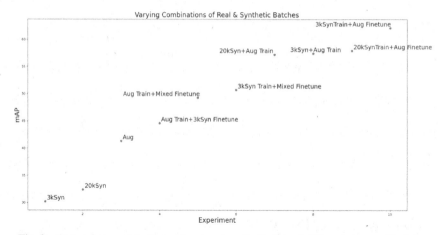

Fig. 9. Comparing combinations of real and synthetic data at different stages of training

for 3K synthetic). Although, as previously validated, the augmented and synthetic data seem to capture different features and, when used in combination, improve over either standalone performance. In the case that we trained with mixed batches from the start (our setting from experiment 1), we reached 57.2 and 57.4 mAP for 20k and 3k synthetic samples. However, our best setting came from pretraining on synthetic and finetuning on real data reaching 62.1 mAP.

4.3 Generalizing to New Tasks

Table 2. Validation mAP for increasing amounts of synthetic data used from each method, starting with the base augmented dataset.

Training data	Syn	Aug	10k Syn + Aug	10k Syn - > Aug
mAP:.50	29.1 ± 2.8	52.9 ± 1.8	63.3 ± 3.0	**66.3 ± 3.9**

Our third experiment looks to test how well our up-sampling approach improves generalization for additional and more complex classes. To that end, we tried to localize the entire relationship indicator bodies and specify two additional gene-gene relationship markers (indirect activate and indirect inhibit). These classes are differentiated from their bases with dashed bodies, as seen in Fig. 8.b.

We annotated the same 250 real diagrams used for the relationship heads, but instead augmented their bodies to 3000 samples. Training with mixed batches again shows considerable improvement over either alone. We also leveraged the insights from experiment 2 and we pretrained on 10,000 synthetic samples with annotated bodies and finetuned on the augmented real samples. These results are listed in Table 2. We can observe that this approach led to a 25% improvement over the model that was just trained with augmented samples.

4.4 Computational Time

We evaluated our computational time by generating 4000 synthetic images on a laptop with 4-Core i5-8265U CPU. The average time per generation of one image with and without multiple threading is 84.5 ms and 67.9 ms, respectively. Even on such an ordinary laptop, the proposed method can generate each synthetic image in a short time. Plus, if we switched to the multi-threading mode, the speedup was 20% higher than that of the single threading mode.

5 Conclusion

This work introduced an up-sampling method for object detection on pathway figures based on a set of rules from spectral space and biological domain knowledge. Such a biology-inspired data augmentation is novel and can better reflect the relationships (gene activation and inhibition). As these relationships are highly diverse in graphics, traditional methods for positional or color modifications do not sample them in depth. Meanwhile, the relationships are represented in specific patterns with biological meanings, GAN-based approaches may not follow the underlying rules well. Our method's fully synthetic approach was able to increase the generalization capacity of the transfer-learned models. Our work motivates further investigation into the upper bound of this synthetic approach and its possible extensions. We also validate the value of a targeted up-sampling approach in addition to traditional augmentation. Our method has the potential for other deep learning data augmentation in research problems where domain knowledge can be represented as rules for generating synthetic data.

Acknowledgements. The research is supported by the National Library of Medicine of the National Institute of Health (NIH) award 5R01LM013392.

References

1. Girshick, R., et al.: Rich feature hierarchies for accu-rate object detection and semantic segmentation. In: Proceedings of the IEEE Conference on Computer Vision and Pattern Recognition (2014)
2. Girshick, R.: Fast R-CNN. In: Proceedings of the IEEE International Conference on Computer Vision (2015)
3. Ren, S., et al.: Faster R-CNN: towards real-time object detection with region proposal networks. IEEE Trans. Pattern Anal. Mach. Intell. **39**(6), 1137–1149 (2016)
4. Lin, T.-Y., et al.: Focal loss for dense object detection. In: Proceedings of the IEEE International Conference on Computer Vision (2017)
5. Lin, T.Y., et al.: Microsoft COCO: common objects in context. In: Fleet, D., Pajdla, T., Schiele, B., Tuytelaars, T. (eds.) ECCV 2014. LNCS, vol. 8693, pp. 740–755. Springer, Cham (2014). https://doi.org/10.1007/978-3-319-10602-1_48
6. Shorten, C., Khoshgoftaar, T.M.: A survey on image data augmentation for deep learning. J. Big Data **6**(1), 1–48 (2019). https://doi.org/10.1186/s40537-019-0197-0
7. Pinceti, A., Sankar, L., Kosut, O.: Synthetic Time-Series Load Data via Conditional Generative Adversarial Networks. arXiv preprint arXiv:2107.03545 (2021)

8. Wang, F., Zhuang, Y., Gu, H., Hu, H.: Automatic generation of synthetic LiDAR point clouds for 3-D data analysis. IEEE Trans. Instrum. Meas. **68**(7), 2671–2673 (2019). https://doi.org/10.1109/TIM.2019.2906416

9. Griffiths, D., Boehm, J.: SynthCity: a large scale synthetic point cloud. arXiv preprint arXiv: 1907.04758 (2019)

10. Kanehisa, M., Goto, S.: KEGG: kyoto encyclopedia of genes and genomes. Nucleic Acids Res. **28**, 27–30 (2000)

11. He, K., et al.: Deep residual learning for image recog-nition. In: Proceedings of the IEEE Conference on Computer Vision and Pattern Recognition (2016)

12. Deng, J., et al.: ImageNet: a large-scale hierarchical image database. In: 2009 IEEE Conference on Com-puter Vision and Pattern Recognition. IEEE (2009)

13. Cooley, J.W., Tukey, J.W.: An algorithm for the machine calculation of complex Fourier series. Math. Comput. **19**(90), 297–301 (1965)

Machine Learning Classifiers Based on Dimensionality Reduction Techniques for the Early Diagnosis of Alzheimer's Disease Using Magnetic Resonance Imaging and Positron Emission Tomography Brain Data

Lilia Lazli[✉] [ID]

Department of Computer and Software Engineering, Polytechnique Montréal,
University of Montreal, Montreal, Canada
Lilia.lazli@polymtl.ca

Abstract. Machine learning techniques have become more attractive and widely used for medical image processing purposes. In particular, the diagnosis of neurodegenerative diseases has recently shown a potential field of application for these methods. The performance comparison of a unique algorithm in various study contexts can be biased, which usually leads to incorrect results. In this context, this study consists in comparing the performance of different machine learning techniques, identifying their main trends and their application for the diagnosis of Alzheimer's disease (AD). We presented a computer-aided diagnosis system for the early diagnosis of AD by analyzing brain data from the OASIS dataset. The principal component analysis (PCA) and the uniform manifold approximation and projection (UMAP) technique have been evaluated on the magnetic resonance imaging and positron emission tomography images as feature selection techniques. After that, the features are fed into nine machine learning models namely Support vector machine (SVM), Artificial neural networks, Decision trees, Random Forests, Discriminant analysis, Regression analysis, Naive Bayes, k-Nearest neighbors, and Ensemble learning. The performance of the proposed classifiers is investigated by the confusion matrix. In addition, area under the curve, Matthews correlation coefficient, accuracy, and F1-score metrics are calculated regarding this matrix. Our results indicate that the SVM-PCA/UMAP schemes provide a significant advantage over the other classifiers. Moreover, they are more efficient than the baseline model based on the voxels-as-features reference feature extraction approach.

Keywords: Alzheimer's disease · Neuroimaging data · Computer-aided diagnosis system · Principal component analysis · Uniform manifold approximation and projection technique · Machine learning classifiers

D. Chicco et al. (Eds.): CIBB 2021, LNBI 13483, pp. 117–131, 2022.
https://doi.org/10.1007/978-3-031-20837-9_10

1 Scientific Background

Alzheimer's disease (AD) is a fatal neurodegenerative disease and one of the world's major public health problems. The WHO estimates that 50 million people are affected by some form of dementia, of which AD is the most common form. It affects 60% to 70% of cases compared to 20% of cases of vascular dementia [1]. The Alzheimer's Disease International 2021 World Report estimates that the annual cost of the AD in the world is approximately one trillion US dollars, which represents the equivalent of more than 1% of world GDP [2].

Although research has revealed a lot about this disease unfortunately, the cause(s) of dementia remain unknown, except for some hereditary forms of the disease. According to researchers from the "Canadian Outcomes Study in Dementia" [3], slowing the progression of the AD will contribute significantly to reducing its economic and psychosocial costs. This slowdown depends on the implementation of early interventions which are possible after an early diagnosis which is crucial to attenuate the effects of AD.

Over the past three decades, computer-aided diagnostic (CAD) systems have become one of the main areas of research in medical imaging and diagnostic radiology [4]. They were introduced with the idea of providing computer output as a "second opinion" to help radiologists assess the extent of disease, and the consistency of radiological diagnosis. The goal is to reduce the false negatives rate and improve diagnostic performance.

With the emergence of data science technology and the electronic processing of medical data, several computational approaches and artificial intelligence techniques are being applied for the purpose of aiding in the diagnosis of diseases. Despite the efforts of researchers, developing an automated AD classification system remains a rather challenging task. Machine learning has offered interesting results in the analysis of medical images; particularly it has shown a prominent result for organ and substructure segmentation, several diseases classification in areas of pathology, brain, breast, bone, retina, etc. [5].

Unfortunately, there is little existing work for AD detection using machine learning models [6–10]. Based on demographic and neuropsychological data, some studies have compared the performance of a few machine learning based models. We cite the work of Kavitha et al. [11] which employed decision tree (DT), random forests (RF), support vector machine (SVM), extreme gradient boosting (XGBoost), and voting classifiers using data from open access series of imaging studies (OASIS). A better validation average accuracy of 83% is obtained with RF and XGBoost. Using the same dataset, Suhaira et al. [12] applied SVM, RF, k-nearest neighbors (KNN) and naïve Bayes (NB) to predict AD from psychological parameters. RF and SVM have achieved the highest degree of accuracy with results that exceed 70%. Williams et al. [13] explore the use of NB, SVM, DT and back-propagation artificial neural network (ANN). Accuracy rates of NB and SVM are significantly higher compared to other models. Amulya et al. [14] applied the gray-level co-occurrence matrix (GLCM) method to extract texture features from OASIS magnetic resonance imaging (MRI) scans and the SVM classifier to discriminate AD patients from healthy control (HC) subjects. The GLCM/SVM classification model provides an average test accuracy of 75.71%. Gray in [15] used the RF classifier to detect AD from multimodal MRI and positron emission tomography (PET) data and she obtains a degree of accuracy of 90.0%. Klöppel et al. in [16] developed a linear SVM based model for AD and HC classification using T1 weighted MRI scan. Up to 96% of AD patients were correctly classified with this model. In [17], Magnin et al. developed SVM based classifier and employed an anatomically labeled brain template to identify regions of interest from whole brain images. Researchers achieved an average of 94.5% correct classification. Several SVM classifiers are developed by Vemuri et al. [18] using genotype and demographic data and structural MRI images to classify AD patients and HC subjects. The highest degree of accuracy was 89.3%. Some research studies [19, 20] have used deep neural Networks for the MRI-based Alzheimer's diagnosis using the OASIS dataset. These methods achieve an accuracy of 93% and 92.39% respectively.

We found in most of the work reported in the literature that researchers provide performance comparisons between different machine learning techniques by referring to studies established by other researchers. However, the same algorithm can provide different results for the same database if the study context, the acquisition and learning parameters, the capacity of the computer equipment, etc. are different.

In other words, comparing the performance of different learning machine algorithms is difficult and can lead to inaccurate results, if these algorithms are used separately and applied in different research. Indeed, the majority of researchers used different parameters and measures for the proposed CAD system citing the type of data used (clinical, demographic, etc.), the cross-validation method followed (10-round validation, 5-round validation, etc. or none), the selection of the type of kernel (linear, polynomial, etc.) for the SVM, the division of each node in the RF as well as the number of trees chosen for the underlying forest associated with each node. Therefore, a performance comparison is reliable between different machine learning techniques, only if a common benchmark on the same database is available and within the reach of researchers.

So, the main purpose of this study is to compare the performance of several machine learning algorithms for early diagnosis of AD. From previous research, it has been proved that MRI and PET scans can perform a significant role for early detection of AD [21]. For this study, we analyze these neuroimaging data for developing multi-modal diagnostic tools, based on principal component analysis (PCA) and uniform manifold approximation and projection (UMAP) approach for feature selection, and multiple machine learning classifiers which are trained to detect AD from MRI and PET neurological images.

The selected classifiers are based on various probabilistic and statistical formalisms as well as optimization concepts. Particularly, we opted for the most efficient techniques namely: SVM, ANN, DT, RF, NB, KNN, discriminant analysis (DA), regression analysis (RA), and ensemble learning (EL). We demonstrate the performance of the CAD system on the OASIS dataset [22] using area under the curve ROC curve (AUC), Matthews correlation coefficient (MCC), accuracy (ACC), and F1-score as quantitative measures [23].

The rest of the paper is organized as follows: Sect. 2 presents our proposed CAD system which combines the advantages of both PCA/UMAP based feature selection, and machine learning based supervised classifiers. Section 3 presents an evaluation of these techniques with experimental details and results described. Finally, we conclude the article in Sect. 4 and present future work.

2 Methods

A generic automated AD classification framework is shown in Fig. 1. In this work, several classifiers such as SVM, ANN, DT, RF, DA, RA, NB, KNN and EL are trained on the PCA/UMAP features extracted from the neurological images from OASIS dataset. We have experienced the performance of these techniques on the T1-weighted MRI and PET scans. In the following subsections, we describe the techniques used in each step of the CAD system. For more details on the mathematical formalism, please refer to [4, 5].

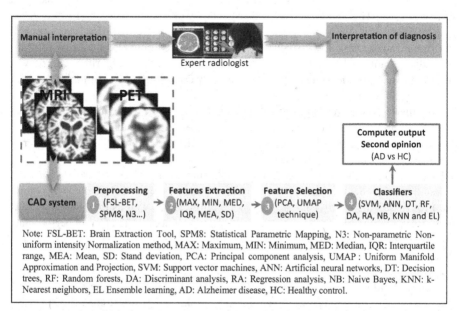

Note: FSL-BET: Brain Extraction Tool, SPM8: Statistical Parametric Mapping, N3: Non-parametric Non-uniform intensity Normalization method, MAX: Maximum, MIN: Minimum, MED: Median, IQR: Interquartile range, MEA: Mean, SD: Stand deviation, PCA: Principal component analysis, UMAP : Uniform Manifold Approximation and Projection, SVM: Support vector machines, ANN: Artificial neural networks, DT: Decision trees, RF: Random forests, DA: Discriminant analysis, RA: Regression analysis, NB: Naive Bayes, KNN: k-Nearest neighbors, EL Ensemble learning, AD: Alzheimer disease, HC: Healthy control.

Fig. 1. Schematic diagram of brain image interpretation (top) and block diagram of corresponding CAD system (bottom). 1) Enhancement of brain image using: - BET-FSL to remove non-brain tissue. - SPM8 model to partially correct the spatial inhomogeneities. - N3 method, to normalize the intensity values. 2) Extraction of voxel features in a mathematical representation using grayscale features as: MAX, MIN, MED, IQR, MEA, and SD. 3) Reduction of voxel parameters by collapsing them into a new feature space applying PCA/UMAP-based feature selection. 4) Classification of images into AD or HC classes using machine learning classifiers such as: SVM, ANN, DT, RF, DA, RA, NB, KNN and EL.

2.1 Dataset Description

We used the OASIS image dataset [22] to evaluate the performance of the proposed CAD system approaches. The images chosen are anatomical MRI and functional PET, and they contain different levels of artifacts (effects of noise, inter-slice intensity variations, and intensity inhomogeneity).

The experiments are performed on images of 300 subjects which include 90 scans of normal subjects and 210 scans of patients with AD ranging from very mild to moderate. Subjects include men and women ages 60 to 96 and for each of them, 3 or 4 T1-weighted MRI scans are available. The MRI data were collected on the Siemens Vision 1.5T scanner (Siemens, Erlangen Germany). The PET data were collected on the Siemens ECAT HRplus 962 PET scanner. Metabolic imaging with [18F] FDG-PET was performed with dynamic 3D acquisition started 40 min after a bolus injection of approximately 5 mCi of FDG and lasted 20 min. For each image $34 \times 47 \times 39$ voxel-sized brain representation is obtained.

2.2 Image Preprocessing

The brain extraction tool FSL-BET [24] was used to remove non-brain tissue in the images. To partially correct the spatial inhomogeneities, we used neuroimaging computer tools, in particular the non-parametric model of the statistical parametric mapping (SPM8) [25] with a minimization function based on the entropy of the image intensity histogram. Moreover, we adapted the Non-parametric Non-uniform intensity Normalization (N3) method [26], to solve the problem of the non-uniform intensity and for normalizing intensity values to ensure the same 0 to 1 dynamic range values for all images.

2.3 Feature Extraction

This process is for creating a new, smaller set of features that stills captures most of the useful information. It can be critical to the success of the proposed feature selection algorithms, and to the convergence of the machine learning algorithms.

First of all, we applied the Voxels-as-features (VAF) method [5] which is the simplest way to create features from an image. We considered these raw voxel values as separate features. In this context, input samples are presented as points in the multidimensional observation space and defined by the measurements of the input features.

Secondly, we calculated the Maximum (MAX), Minimum (MIN), Median (MED), Interquartile range (IQR), Mean (MEA), and Stand deviation (SD) values which represents the most relevant grayscale features offering the best performance for this search compared to other parameters tested previously. These features are extracted for the hippocampal and parietal cortices which are the parts of the brain most affected by AD and from which we have observed a large reduction in brain volume.

2.4 Dimensionality Reduction Techniques

Working in high-dimensional spaces is undesirable for many reasons; data analysis is usually computationally intractable. For this reason, we have applied the most common techniques of dimensionality reduction, which allows the transformation of data from a high-dimensional space into a low-dimensional space with a representation that retains some significant properties of the original data.

Principal Component Analysis. The PCA consists in transforming correlated variables between them into new variables uncorrelated from each other that successively maximize variance and which represent the principal components or axes [27]. In this research, the voxels of each sample are rearranged into a vector form (MAX, MIN, MED, IQR, MEA, and SD). With using the specific PCA steps, 1) we standardize the original data (all the samples are subtracted from the mean value of corresponding feature); 2) we calculate the covariance matrix, the eigenvalue and the eigenvector; 3) we record the resulting eigenvalues in the order of large to small, and calculate the contribution rate of each principal component; and 5) we transform the original sample matrix into a new matrix.

Uniform Manifold Approximation and Projection Approach. The UMAP is based on manifold learning techniques and ideas from topological data analysis [28]. In the first phase of the UMAP algorithm, we use Nearest-Neighbor-Descent algorithm which consists in constructing a fuzzy topological representation for which neighborhood graph based approach should capture the structure of the manifold during dimension reduction. The second phase, we use stochastic gradient descent to optimize the low-dimensional representation which allows to have a fuzzy topological representation as close as possible as measured by the cross-entropy.

2.5 Machine Learning Classifiers

We applied techniques based on supervised machine learning that takes into account labeled data from different patients in the learning phase. At first, the training data are used to train the model and readjust its parameters progressively until convergence. Then, the model with the final parameters is applied on unlabeled test data for the classification of the unknown subjects into similar classes, the class of AD patients and that of HC subjects.

Support Vector Machines. The SVMs try to maximize the margin between classes by finding the optimal values, in the quadratic programming problem (represented in dual Lagrangian form) [21]. Unlabelled instances are classified using the learned parameters and bias, by taking the sign of the appropriate decision function. SVMs separate binary labeled training data, by the hyperplane. This hyperplane is maximally distant from the two classes (known as the maximal margin hyperplane). When no linear separation of the training data is possible, SVMs can work effectively in combination with kernel techniques. Thus, the hyperplane defining the SVMs corresponds to a nonlinear decision boundary in the input space.

Artificial Neural Network. The ANN can be viewed as weighted directed graphs in which artificial neurons are nodes and directed edges (with weights) are connections between neuron outputs and neuron inputs [10]. The development of the backpropagation learning algorithm for determining weights in a multilayer perceptron has made these networks the most popular among ANN researchers. For our experiments a feed-forward neural network (FFNN) was used. This type of ANN is static, that is, it produces only one set of output values rather than a sequence of values from a given input. Because the

FFNN is memory-less, its response to an input is independent of the previous network state.

Discriminant Analysis. Linear discriminant analysis (LDA) is the most used DA technique [29]. LDA allows data to be projected into a moderate dimensional feature space with a true class of separable features that minimizes computational costs and overfitting. We followed the following three-step process to project features onto a lower dimensional space: 1) Calculate the distance between the mean of the two classes (between-class variance). 2) Calculate the distance between the mean and each datum of each class (within-class variance). 3) Create the lower dimensional space that maximizes between-class variance and minimizes within-class variance.

Regression Analysis. The RA is among the predictive modeling approaches that allows the study of the relationship between a dependent variable (target) and one or more independent variables (predictor). We applied the logistic regression (LR) technique [30] which allows calculating the probability of mutually exclusive occurrences (0/1 in our case). In this context, the target variable can take only one of these two values. A sigmoid curve is usually drawn to represent the connection of the target variable to the independent variable, associating a probability with a value between 0 and 1. To assign a new datum to one of the two classes (AD/HC), this technique calculates the probability of the instance belonging to each class. To apply the LR as a binary classifier, it is therefore necessary to assign a threshold to distinguish between the two classes. In our case, a probability value greater than 0.50 makes it possible to assign the input data to class AD if not to class HC.

Naive Bayes. The naive Bayes classifier is based on Bayes' theorem with the application of strong (naive) independence assumptions between the features [31]. This classifier is highly scalable, which requires a number of linear parameters for the learning step. Generally, the parameter estimation is based on maximum likelihood. Maximum likelihood training is achieved by evaluating a closed-form expression, which requires linear time, rather than an expensive iterative approximation like that employed for many other types of classifiers. The advantage of this classifier is that it requires relatively little training data to estimate the parameters necessary for the classification, namely means and variances of the different variables. So, we are interested in calculating the posterior probability $P(h|d)$ from the prior probability $P(h)$ with $P(d)$ and $P(d|h)$. After calculating this probability for several different hypotheses, the hypothesis with the highest probability is selected which represents the maximum a posteriori (MAP) hypothesis.

k-Nearest Neighbors. The KNN algorithm represents the simplified version of NB, except that KNN does not need to consider probability values [32]. Generally, we have a learning database made up of N "input-output" pairs. To estimate the output associated with a new input x, the KNN method consists of considering (identically) the k learning samples whose input is closest to the new input x, according to a distance to be defined. The training examples are vectors in a multidimensional feature space, each with a membership class label. The learning phase of the algorithm consists only in storing the feature vectors and the class labels of the learning samples. In the classification phase, k is a user-defined constant, and an unlabeled vector (test point) is classified by assigning it the label that is most frequent among the k training samples closest to the point to be classified.

Decision Trees. The DTs are a rule-based approach that can perform both classification and regression tasks. They represent a tree structure of several levels presented by nodes whose highest node represents the root node. The goal is to create a model that predicts the value of a target variable given several input variables. So, DTs model the decision logic to process the data, perform the test, and match the result to classify the dataset in the tree. The tests are represented by the intermediate nodes (nodes having at least one child node) and they are carried out from the variables associated with the input (attributes). The result obtained from the current test allows branching to the appropriate child node, where the process is repeated until reaching the leaf node located in the last level of the tree structure [33]. In other word, the algorithm tries to completely separate the dataset so that all leaf nodes belong to a single class. On every split, the classification algorithm tries to divide the dataset into the smallest subset possible. So, like any other machine learning algorithm, the goal is to minimize the loss function (Stochastic Gradient Descent in our case) as much as possible.

Random Forests. The RFs are a meta estimator that fits a number of DTs classifiers on various subsamples of the dataset and uses averaging to improve the predictive accuracy and control over-fitting [34]. The RFs are based on two key concepts that give it the name random: A random sampling of training data set when building trees and random subsets of features considered when splitting nodes. In this study, a bagging technique is used to create an ensemble of trees where multiple training sets are generated with replacement. We divided a dataset into N samples using randomized sampling. Then, during the training phase, the DTs of forest are used to train different parts of the training data. While each test data is transmitted to the different DTs in the forest, each of them provides its classification result. The concept of voting is considered by the forest to choose the most appropriate result.

Ensemble Learning. The EL is a way to generate various base classifiers from which a new classifier is derived that performs better than any constituent classifier [35]. For independently constructing ensembles, we used in this study, the RF which is an extension on bagging. Each classifier in the ensemble is a DT classifier and is generated using a random selection of attributes at each node to determine the split. During classification, each tree votes and the most popular class is returned. We followed the following steps for the construction of the classifier. 1) Multiple subsets are created from the original data set, selecting observations with replacement. 2) A subset of features is randomly selected and the feature that gives the best split is used to iteratively split the node. 3) The tree is grown to the largest. 4) Repeat the above steps and the prediction is given based on the aggregation of predictions from n number of trees.

2.6 Description of resampling Method and Performance Metrics

Each subject is represented by grayscale features which are MAX, MIN, MED, IQR, MEA, and SD, and is collapsed into a new feature space by applying the feature selection based on PCA and UMAP technique. The 10-fold cross-validation procedure is used to test the performance of the classifiers and its robustness. For each fold, we have used 70% as training data and 30% as test data which is used to forecast results or to assess the correctness of our models. To analyze the performance of the proposed classifiers, for the first factor, the confusion matrix is generated for each classifier, and regarding this confusion matrix, some quantitative measures which are ACC, MCC, F1-Score, and AUC are calculated.

In our diagnostic problem the training datasets suffer from imbalance of targeted data. This is due to artifacts of the images. Typically, this data heavily imbalanced affect performance evaluation of trained models. For this reason, in addition to calculating the classification accuracy, we estimated the MCC, AUC and F1-Score.

The confusion matrix based on four categories (true positives, false negatives, true negatives, and false positives) is used to evaluate a classifier's performance considering a pre-known set of labeled data. The sensitivity or recall metric shows the likelihood of predicting true positive, while the specificity measures the true negative rate. The F1-Score shows the balance between sensitivity and precision. The accuracy of each model is measured by evaluating the trueness of the results. The MCC is a more reliable statistical rate which produces a high score only if the prediction obtained good results in all of the four confusion matrix categories. The ROC compares the classifiers' performance among the whole range of class distributions and error costs. To compare the ROC curves, the AUC is calculated.

3 Result and Discussion

Figure 3 presents the confusion matrix of all the classifiers for MRI images. Again, SVM outperforms other models for all images. The sensitivity is higher than those of other classifiers, proving the power of these machines to better identify true positives. Moreover, Fig. 2. presents the ROC curve for the SVM classifier based on the Gaussian kernel function which provided the best results compared to the other classifiers.

From Table 1 which illustrates the results concerning the performance of the different machine learning classifiers, it is observed that the SVM-based CAD system provided superior results for both processed images, with prediction speed of ~1500 obs/sec and training time of 1,8204 s. The ANN comes second considering the results of the other classifiers. It consumed more time for training with time of 8,0284 s and prediction speed of ~4000 obs/sec.

For dimensional reduction techniques, from the Table 1, we found that the PCA and UMAP offer almost the same performance and exceed that of VAF-based systems. Moreover, in all cases and for both PET and MRI datasets, the Median & IQR combination based feature selection offered the best PCA results compared to the other combinations (MEA & SD, MIN & MAX, etc.). This finding is perhaps because the samples distribution is skewed and not symmetric and there are extreme outliers present whose interquartile range is the best way to measure the dispersion of this type of data.

In terms of computation time the Fig. 4 shows the results for the proposed PCA and UMAP techniques compared to the t-distributed stochastic neighbor embedding (t-SNE) technique. We see the advantages of the PCA and UMAP over t-SNE. Although UMAP is slower than PCA, it is clearly the next best performance option among the implementations explored here. Of this effect, and given the quality of the results that UMAP has provided, we believe that it is also a good approach for dimension reduction. It is has a solid theoretical backing but the main limitation of UMAP is its lack of maturity. It is a very new technique, so the libraries and best practices are not yet firmly established or robust.

In general, the proposed CAD system based on PCA/UMAP feature selection and SVM/ANN classifiers can detect early AD and successfully classify the major AD patients and discriminate them from HC subjects. These preliminary results of evaluating the complete CAD system are shown to be more useful for separating NORMAL and AD classes.

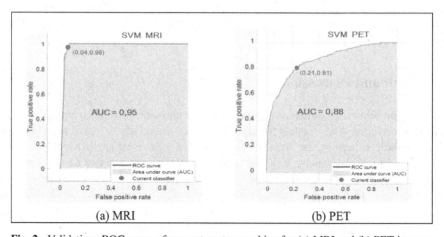

Fig. 2. Validations ROC curve of support vector machine for (a) MRI and (b) PET images.

Table 1. Classification performance in term of ACC, MCC, F1-SCORE and AUC obtained from the evaluation of several machine learning classifiers using PCA and UMAP based feature selection and 10-fold cross-validation.

	ACC (%)						F1-SCORE					
	MRI			PET			MRI			PET		
	VAF	PCA	UMAP	VAF	PCA	UMAP	VAF	PCA	UMAP	VAF	PCA	UMAP
SVM	62.45	**91.06**	**89.18**	66.04	**90.47**	90.04	0.48	**0.80**	**0.79**	0.61	**0.88**	**0.85**
ANN	60.52	**88.32**	**85.12**	65.01	86.97	84.15	0.45	**0.78**	**0.78**	0.59	**0.87**	**0.85**
DT	52.03	86.23	83.79	64.02	84.14	84.00	0.39	0.77	0.75	0.39	0.80	0.80
RF	51.42	80.04	80.15	54.12	82.17	79.14	0.42	0.76	0.71	0.46	0.82	0.81
DA	60.78	79.04	79.00	60.78	80.54	78.23	0.44	**0.78**	0.70	0.36	0.77	0.75
RA	45.18	72.29	72.11	59.64	79.07	78.89	0.40	0.76	0.70	0.45	0.79	0.76
NB	60.09	86.82	86.00	61.07	84.36	82.99	0.40	0.71	0.70	0.33	0.80	0.75
KNN	59.54	85.99	84.75	61.08	83.95	80.47	0.39	0.73	0.69	0.57	0.80	0.80
EL	61.89	87.01	86.02	64.36	84.04	83.55	0.45	**0.78**	**0.78**	0.57	0.86	0.84

	MCC						AUC					
	MRI			PET			MRI			PET		
	VAF	PCA	UMAP	VAF	PCA	UMAP	VAF	PCA	UMAP	**VAF**	PCA	UMAP
SVM	0.60	**0.96**	**0.92**	0.39	**0.95**	**0.93**	0.62	**0.95**	**0.87**	0.63	**0.88**	**0.87**
ANN	0.58	**0.95**	**0.90**	0.38	**0.95**	**0.90**	0.58	**0.85**	**0.86**	0.61	**0.84**	**0.83**
DT	0.48	0.75	0.73	0.32	0.84	0.82	0.51	0.84	0.80	0.60	0.82	0.81
RF	0.58	0.84	0.81	0.30	0.79	0.80	0.49	0.77	0.74	0.61	0.80	0.77
DA	0.47	0.80	0.80	0.29	0.82	0.82	0.60	0.77	0.72	0.58	0.77	0.78
RA	0.41	0.79	0.77	0.29	0.76	0.74	0.43	0.71	0.77	0.54	0.78	0.79
NB	0.47	0.88	0.79	0.32	0.87	0.73	0.57	0.84	0.75	0.59	0.82	0.81
KNN	0.59	0.79	0.74	0.32	0.82	0.72	0.57	0.83	0.80	0.54	0.81	0.80
EL	0.58	0.91	0.86	0.35	0.90	**0.90**	0.58	**0.85**	0.83	0.61	0.83	0.80

Note (Hyper-parameters). DT: Maximum number of splits = 100; KNN: Number of neighbors = 10; SVM: Box constraint level = 1, Kernel scale = 2.8; EL: Number of learners = 30, Maximum number of splits = 121, Learning rate = 0.1; ANN: Number of fully connected layers = 2, First layer size = 10, Second layer size = 10, Activation = ReLU, Iteration limit = 1000, Regularization strength (Lambda) = 0; RF: max depth = 120, min sample split = 3.

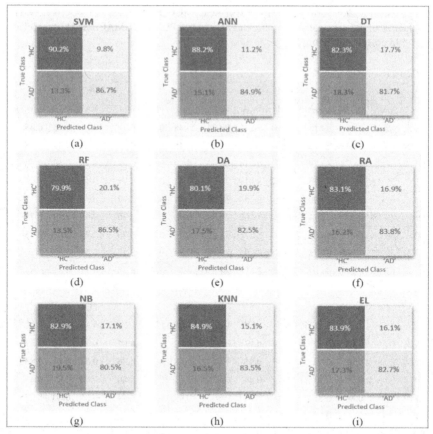

Fig. 3. Confusion matrix of MRI images, (a) Support vector machine, (b) Artificial neural networks, (c) Decision trees, (d) Random Forests, (e) Discriminant analysis, (f) Regression analysis, (g) Naive Bayes, (h) K-Nearest neighbors, (i) Ensemble learning.

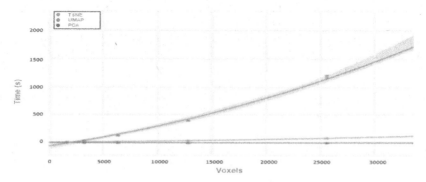

Fig. 4. Computation time of PCA, UMAP and t-SNE based feature selection.

4 Conclusion

In this study, we proposed a CAD system to diagnose patients with probable prognosis of AD. Various schemes combining different feature extraction and selection techniques with several classification models have been studied and tested on brain database to distinguish between brain images belonging to AD patients and those of normally aging subjects. The PCA and UMAP based feature selection techniques have been evaluated on the MRI/PET database. After that, the features are fed into nine classifiers based on supervised machine learning namely SVM, ANN, DT, RF, DA, RA, NB, KNN and EL.

With using 10 cross-validation strategy and ACC, MCC, F1-SCORE and AUC as quantitative measures for the classification performance, all experimental results obtained showed that this kind of machine learning classifiers represent a promising approach in the presence of incomplete and imprecise training knowledge, thus allowing flexible adaptation of the classification architecture to the available information. Particularly, the PCA/UMAP-SVM scheme works in general better than other classifiers, yielding in most cases higher performance rates.

We conclude that the experiments reported in this paper indicate that artificial intelligence techniques can be used to automate aspects of clinical diagnosis of individuals with cognitive impairment, which may have significant implications for the future of health care. So, the CAD system can improve doctors' diagnostic capabilities and reduce the time required for accurate diagnosis.

In the future, we are exploring semi-supervised systems and unsupervised deep learning methods for multi-class diagnosis of AD. In addition, data extensive knowledge and experience are required to well distinguish the AD data from the aged normal. Research is in progress integrating other sequences types include T2-weighted, FLAIR, and TSE for MRI images and PiB and AV45 for PET images. We also plan to integrate other type of features such as demographics data (gender, age, Right Handed, APOE, Race), neuropsychological data (memory, language, I.Q., visual-spatial skills), genetic and biospecimen data.

Acknowledgments. The author would like to thank the FRQNT organization for its financial support offered to accomplish this project. Many thanks to the researchers and expert clinicians of the OASIS dataset, for developing the images used in the preparation of this work. Special thanks to the reviewers of this work.

Funding. This project (#2022-2023-B3X-314498) was funded by the Fonds de recherche du Québec-Nature et Technologies (FRQNT).

References

1. OMS. https://www.who.int/fr/news-room/fact-sheets/detail/dementia. Accessed 16 Nov 2021
2. Organization Alzheimer's Disease International. https://www.alzint.org/u/World-Alzheimer-Report-2021.pdf. Accessed 16 Nov 2021
3. Herrmann, N., Harimoto, T., Balshaw, R., Lanctôt, K.L.: Canadian outcomes study in dementia (COSID) investigators: risk factors for progression of Alzheimer disease in a Canadian population: the Canadian outcomes study in dementia (COSID). Canadian journal of psychiatry. Revue canadienne de psychiatrie **60**(4), 189–199 (2015)

4. Salem, H., Soria, D., Lund, J.N., et al.: A systematic review of the applications of expert systems (ES) and machine learning (ML) in clinical urology. BMC Med. Inform. Decis. Mak. **21**, 223 (2021)

5. Lazli, L., Boukadoum, M., Aït-Mohamed, O.: A survey on computer-aided diagnosis of brain disorders through MRI based on machine learning and data mining methodologies with an emphasis on Alzheimer disease diagnosis and the contribution of the multimodal fusion. Appl. Sci. **10**(5) (2020)

6. Biagetti, G., Crippa, P., Falaschetti, L., Luzzi, S., Santarelli, R., Turchetti, C.: Classification of Alzheimer's disease from structural magnetic resonance imaging using particle-bernstein polynomials algorithm. In: Czarnowski, I., Howlett, R.J., Jain, L.C. (eds.) Intelligent Decision Technologies 2019. SIST, vol. 143, pp. 49–62. Springer, Singapore (2019). https://doi.org/10.1007/978-981-13-8303-8_5

7. Karami, V., Nittari, G., Amenta, F.: Neuroimaging computer-aided diagnosis systems for Alzheimer's disease. Int. J. Imaging Syst. Technol. **29**(1), 83–94 (2019)

8. Lazli, L., Boukadoum, M., Aït-Mohamed, O.: Computer-aided diagnosis system for Alzheimer's disease using fuzzy-possibilistic tissue segmentation and SVM classification. In: 2018 IEEE Life Sciences Conference (LSC), Montréal, Canada, pp. 33–36 (2018)

9. Alvarez Fernandez, I., Aguilar, M., González, et al.: Clinic-based validation of cerebrospinal fluid biomarkers with florbetapir PET for diagnosis of dementia. J. Alzheimer's Dis. **61**, 1 9 (2018)

10. Cheng, D., Liu, M.: CNNs based multi-modality classification for AD diagnosis. In: Proceedings of the 10th International Congress on Image and Signal Processing, Bio Medical Engineering and Informatics (CISP-BMEI), Shanghai, China, pp. 1–5 (2017)

11. Kavitha, C., Vinodhini, M., Srividhya, S.R., et al.: Early-stage Alzheimer's disease prediction using machine learning models. Front. Public Health **10** (2022)

12. Suhaira, V.P., Sita, S., Joby, G.: Alzheimer's disease: classification and detection using MRI dataset. Int. J. Innov. Technol. Explor. Eng. **10**(5) (2021)

13. Williams, J.A., Weakley, A., Cook, D.J., et al.: Machine learning techniques for diagnostic differentiation of mild cognitive impairment and dementia. Assoc. Adv. Artif. Intell. (2013). https://www.researchgate.net/publication/286817504

14. Amulya, E.R., Varma, S., Paul, V.: Classification of brain MR images using texture feature extraction. Int. J. Comput. Sci. Eng. **5**(5), 1722–1729 (2017)

15. Gray, K.R.: Machine learning for image-based classification of Alzheimer's disease. Ph.D. thesis, Imperial College London (2012)

16. Klöppel, S., Stonnington, C.M., Chu, C., et al.: Automatic classification of MR scans in Alzheimer's disease. Brain **131**(3), 681–689 (2008)

17. Magnin, B., Mesrob, L., Kinkingnéhun, S., et al.: Support vector machine-based classification of Alzheimer's disease from whole-brain anatomical MRI. Neuroradiology **51**(2), 73–83 (2009)

18. Vemuri, P., Gunter, J.L., Senjem, M.L., et al.: Alzheimer's disease diagnosis in individual subjects using structural MR images: validation studies. Neuroimage **39**(3), 1186–1197 (2008)

19. Saratxaga, C.L., Moya, I., Picón, A., et al.: MRI Deep learning-based solution for Alzheimer's disease prediction. J. Pers. Med. **11**, 902 (2021)

20. Basheer, S., Bhatia, S., Sakri, S.B.: Computational modeling of dementia prediction using deep neural network: analysis on OASIS dataset. IEEE Access **9**, 42449–42462 (2021)

21. Lazli, L., Boukadoum, M., Ait-Mohamed, O.: Computer-aided diagnosis system of Alzheimer's disease based on multimodal fusion: tissue quantification based on the hybrid fuzzy-genetic-possibilistic model and discriminative classification based on the SVDD model. Brain Sci. **9**(10), 289 (2019)

22. Marcus, D.S., Wang, T.H., Parker, J., et al.: Open access series of imaging studies (OASIS): cross-sectional MRI data in young, middle aged, nondemented, and demented older adults. J. Cognit. Neurosci **19**(9), 1498–1507 (2007)

23. Chicco, D., Giuseppe, J.: The advantages of the Matthews correlation coefficient (MCC) over F1 score and accuracy in binary classification evaluation. BMC Genomics **21**(1), 6 (2020)

24. Smith, S.M.: Fast robust automated brain extraction. Hum. Brain Mapp. **17**, 143–155 (2002)

25. Statistical parametric mapping tool. https://www.fil.ion.ucl.ac.uk/spm/software/spm

26. N3 software. http://www.bic.mni.mcgill.ca/software/N3/. Accessed 17 Oct 2020

27. Jolliffe, I.T., Cadima, J.: Principal component analysis: a review and recent developments. Philos. Trans. R. Soc. Math. Phys. Eng. Sci. **374**(2065) (2016)

28. Kwok, H., Lai Guan, N., Ginhoux, F., Newell, E.W.: Dimensionality reduction for visualizing single-cell data using UMAP. Nat. Biotechnol. **37**(1), 38 44 (2019)

29. Tharwat, A., Gaber, T., Abdelhameed, I., Aboul Ella H.: Linear discriminant analysis: a detailed tutorial. AI Commun. **30**, 169–190 (2017)

30. Hosmer Jr., DW., Lemeshow, S., Sturdivant, RX.: Applied Logistic Regression. Wiley, New York (2013)

31. Rish, I.: An empirical study of the naive Bayes classifier. In: IJCAI 2001 Workshop on Empirical Methods in Artificial Intelligence, vol. 3, no. 22, pp. 41–46. IBM, New York (2001)

32. Cover, T., Hart, P.: Nearest neighbor pattern classification. IEEE Trans. Inf. Theory. **13**(1), 21–27 (1967)

33. Quinlan, J.R.: Induction of decision trees. Mach. Learn. **1**(1), 81–106 (1986)

34. Breiman, L.: Random forests. Mach. Learn. **45**(1), 5–32 (2001)

35. Ayerdi, B., Savio, A., Graña, M.: Meta-ensembles of classifiers for Alzheimer's disease detection using independent ROI features. In: Ferrández Vicente, J.M., Álvarez Sánchez, J.R., de la Paz López, F., Toledo Moreo, F.J. (eds.) IWINAC 2013. LNCS, vol. 7931, pp. 122–130. Springer, Heidelberg (2013). https://doi.org/10.1007/978-3-642-38622-0_13

Text Mining Enhancements for Image Recognition of Gene Names and Gene Relations

Yijie Ren[1], Fei He[2], Jing Qu[2], Yifan Li[2], Joshua Thompson[1], Mark Hannink[1], Mihail Popescu[1], and Dong Xu[1(✉)]

[1] University of Missouri, Columbia, MO 65211, USA
jltmh3@umsystem.edu, {hanninkm,xudong}@missouri.edu,
popescum@health.missouri.edu
[2] Northeast Normal University, No. 2555 Jingyue Street, Changchun 130117, China
{hef740,quj165,liyf994}@nenu.edu.cn

Abstract. The volume of the biological literature has been increasing fast, which leads to a rapid growth of biological pathway figures included in the related biological papers. Each pathway figure encompasses rich biological information, consisting of gene names and gene relations. However, manual curations for pathway figures require tremendous time and labor. While leveraging advanced image understanding models may accelerate the process of curations, the accuracy of these models still needs improvements. Since each pathway figure is associated with a paper, most of the gene names and gene relations in a pathway figure also appear in the related paper text, where we can utilize text mining to improve the image recognition results. In this paper, we applied a fuzzy match method to detect gene names with different "gene dictionaries," as well as gene co-occurrence in the plain text for suggesting gene relations. We have demonstrated that the performance of image understanding for both gene name recognitions and gene relation extractions can be improved with the help of text mining methods. All the data and code are available at GitHub (https://github.com/lyfer233/Text-Mining-Enhancements-for-Image-Recognition-of-Gene-Names-and-Gene-Relations).

Keywords: Text mining · Biological pathway · Gene name · Gene relation

1 Introduction

Biological pathway figures in research papers describe the process of a biological function, which involves many genes and their interactions. The gene names and their relations in pathway figures are great resources for biological studies and downstream applications. As the biological literature grows fast, the volume of pathway figures increases fast as well. The manual curations of the gene names and the corresponding relations in pathway figures are time-consuming and labor-consuming, which cannot catch up with the rapid growth of publications. Therefore, automatic information extraction from pathway figures is necessary to reduce the workload of manual curations, which help identify both gene names and gene relations in the figures.

D. Chicco et al. (Eds.): CIBB 2021, LNBI 13483, pp. 132–142, 2022.
https://doi.org/10.1007/978-3-031-20837-9_11

Our recent work [1] annotated gene regulatory relations using image recognition can be complemented by the text-mining approach, where both gene names and gene relations can be recognized from plain biomedical text. This study combined both gene name recognition and gene relation recognition approaches to enhance gene-relationship recognition from biomedical text. In particular, we used different gene name dictionaries to apply exact match and fuzzy match to help increase the accuracy of image recognition results. Meanwhile, gene-relation information can be derived from gene co-occurrences in the text, from both the same sentence and neighbor sentences, with three different levels, namely caption level, paragraph level, and full-text level. Finally, we compared several methods to enhance the image recognition results of gene names and gene relations from pathway figures.

From the results we provided in Sect. 4, we believe our text mining methods will offer combination of pipelines to help improve the performance of pathway image recognitions. At the same time, we have done a thorough analysis to explain why some methods did not perform well based on our experiments. The contributions of this paper are as follows:

- To the best of our knowledge, this is the first study combining image understanding and natural language processing for recognizing gene names and relationships from the biomedical literature.
- We utilized fuzzy matches (introduced in Sect. 3.3) to improve the image recognition results of gene names. Particularly, we applied various "gene dictionaries" in fuzzy matches to obtain enhancement results.
- We adopted and compared gene co-occurrence at different levels (introduced in Sect. 3.4) as different gene relation enhancement methods, demonstrating that gene relation enhancements from the text could improve the image recognition results of gene relations.

2   Related Work

Gene name identifications from pathway figures [2] provide a basis for pathway figure information extraction, which can be useful for updating pathway databases such as KEGG [3]. However, gene name recognitions based on image recognition tools often contain errors. To solve this problem, we adopted text mining methods to enhance image recognition results. Many text mining tools use Named Entity Recognition (NER) as the core method to recognize named entities in the text. The same technique could also be applied to the biological literature, which identifies biomedical concepts from the text, such as PubTator Central [4]. BioBERT [5] is also a good pretrained language model for NER tasks, and BERN [6], a BioBERT-based biomedical text mining tool provides a similar web service as PubTator Central to recognize known biomedical entities.

Various biological relation extraction methods, such as the node2vec model [7], have been applied to extract different types of biological relations from papers. They extract different relation types, such as gene-gene relationships and gene-disease associations [8]. However, they usually extract biological relations solely from the text, while we are trying to enhance the gene relation results from image recognition.

3   Methods

3.1   Dataset

We retrieved and manually curated 45 pathway figures containing 311 gene names and 193 gene relations in total for analyzing the contributions of the proposed text mining methods. Our text mining enhancements complement the results of our recent image recognition pipeline [1], which integrates both the deep learning model and the optical character recognition (OCR) as Sect. 3.2 described. The manual curation results are treated as the ground truth when conducting the result comparison in Sect. 3. There are many types of gene relations in pathway figures, but in this study, we only focus on two types of gene regulatory relations – "inhibit" and "activate".

3.2   OCR Tool

OCR serves as a technique to recognize the text within an image. We leverage Google's Could Vision API [9] to extract text from pathway figures. It can process entire images directly without the need for additional preprocessing (e.g., deskewing, resizing, etc.) by simply calling its built-in text-detection method and passing the current figure and a suggested language. The API response includes all the words found in the figures and a hierarchical breakdown specifying pages, blocks, paragraphs, words, and symbols from the text. By extracting the words in each paragraph, we can build phrases that appear frequently to enhance gene name and relation recognition in our figures.

3.3   Gene Name Enhancements

After obtaining the image recognition results, we first compared the recognized gene names with a certain gene dictionary to see if an exact match was found. There are four dictionaries in our attempts – manual curations, whole article, PubTator Central [4], and exHUGO, where the dictionary of manual curations is served as a "control group". The dictionary of manual curations is the ground truth that we curated each pathway figure by ourselves. The dictionary of the whole article is the full text of the article containing the pathway figures, where we tokenized each word inside the full text as an entry in the dictionary.

PubTator Central is a deep learning-based annotation tool for biological concepts, such as genes and chemicals, as illustrated in Fig. 1. All the gene annotations extracted by its deep learning model are stored in the database, and we can easily retrieve the gene annotations of an article via its Web API.

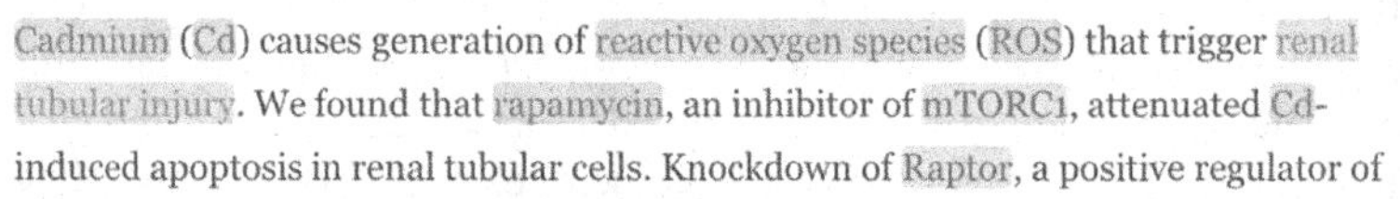

Cadmium (Cd) causes generation of reactive oxygen species (ROS) that trigger renal tubular injury. We found that rapamycin, an inhibitor of mTORC1, attenuated Cd-induced apoptosis in renal tubular cells. Knockdown of Raptor, a positive regulator of

Fig. 1. An example of biological concept annotation from PubTator Central [4]. It contains various biological concepts marked by different colors, where we only focus on the genes, colored by purple here. (Color figure online)

The dictionary of exHUGO combines the HUGO (Human Genome Organisation) gene name list [10] and our manual curations. Customized from HGNC (HUGO Gene Nomenclature Committee), HUGO gene list contains all the approved gene names recorded inside their database. The HUGO gene list only includes regular gene names of humans; however, biologists usually adopted their idiomatic expressions for genes in plotting pathway figures. Therefore, we extended the HUGO gene list with some manually curated gene symbols in a commonly used way forming the exHUGO dictionary.

If there is no exact match found for the gene names between the results of the image recognition pipeline and the dictionaries, we applied a fuzzy match (Levenshtein distance [11]) to the gene names. Levenshtein distance is a method to measure the differences between two strings – a and b, which can indicate how many characters are needed to change from one word to another, as Eq. 1 shows:

$$
Lev(a, b) = \begin{cases} |a| & if\ |b| = 0 \\ |b| & if\ |a| = 0 \\ lev(tail(a), tail(b)) & if\ a[0] = b[0] \\ 1 + min \begin{cases} lev(a, tail(b)) \\ lev(tail(a), tail(b)) \\ lev(tail(a), b) \end{cases} otherwise \end{cases}
\tag{1}
$$

where the tail means the remaining string after its first character, and a[0] and b[0] represent the first character of string a and string b, respectively. For each OCR result, we compare it with all gene names in the dictionary and pick the one with the largest

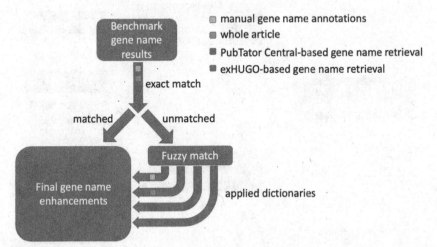

Fig. 2. The process of gene name enhancements. After obtaining the benchmark gene name results, we first compare them with the four different dictionaries, marked by the 4 different colors in the legend. If there are direct matches, these gene names become part of the final gene name results. If not, we then apply fuzzy match to these gene names through the same 4 dictionaries, to discover more gene names. Together with the direct matches, they become the complete final gene name results.

similarity as its possibly corrected gene name. If the similarity is larger than a defined threshold (0.9 is set in this study), the matched gene name will replace the OCR result. Otherwise, the OCR result is considered no matched gene and treated as non-gene text.

We then applied the same gene dictionaries to fuzzy matches we used for the exact match during the first step. Figure 2 shows the process of the whole gene name enhancements.

3.4 Gene Relation Enhancements

After obtaining the gene relation results from our image recognition pipeline [1], we applied two different gene relation extraction methods within three different text region levels to improve the image recognition results. The two different gene relation extraction methods refer to the gene co-occurrence in the same sentence and in neighboring sentences, whereas the three different levels correspond to caption level, paragraph level, and full-text level, as shown in Fig. 3. Intuitively, the caption level is the caption of a specific pathway figure, while the paragraph level is the paragraph that mentions the pathway figure. Full-text level is all the text of the paper containing the pathway figure.

Gene co-occurrence is computed by how many times two genes co-occur in the selected text region. We calculated all gene name pairs recognized in the pathway figures in both the same sentence and neighboring three sentences before and after a specific sentence, which adds to seven sentences in total. For both gene co-occurrences in

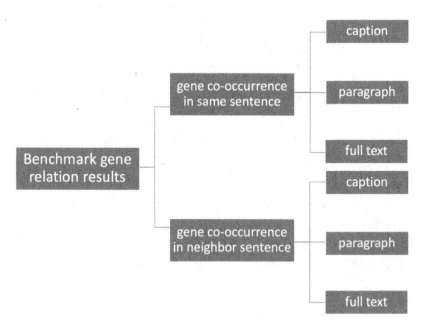

Fig. 3. The process of gene relation enhancements. After obtaining the benchmark gene relation results, we measure the gene co-occurrence in 2 ways – both the same sentence and neighbor 3 sentences, turning out to 7 sentences in total. For each way, we count the gene co-occurrences in 3 different levels – caption level, paragraph level and full-text level.

the same sentences and neighboring sentences, we calculated in three different levels, as mentioned above. Since gene relation recognitions always happen after gene name recognitions in the image recognition pipeline, we assume all gene names in the pathway figures are recognized correctly before improving gene-relation recognition. Similarly, the direct gene relation results from image recognition are also corrected by gene names to conduct a fair comparison with other enhancements.

4 Results

4.1 Gene Name Enhancement Results

The number of annotated gene names in 45 pathway figures ranges from 2 to 20. We listed the average of precisions, recalls, and F1 scores after enhancements for all four dictionaries when employing fuzzy match, together with the corrected original image recognition results, in Table 1. The "Image only" column indicates the initial results from image recognition, and the other four columns show the results after applying a fuzzy match with different dictionaries. The column "Manual curations" means that the dictionary applied to fuzzy match is the gene names manually curated from each pathway figure, which can be treated as the "control group", since the manual curations are not always available for pathway figures. The column "Whole article" utilizes the whole article to be the gene dictionary for a fuzzy match, where the whole text of a paper is tokenized into single words as entries in the dictionary. The column "PubTator Central received" treats the gene names retrieved from PubTator Central as the gene dictionary, while the column "exHUGO retrieved" is based on exHUGO dictionary.

The comparison results show improvements by all three fuzzy match dictionaries. The reason for relatively low performance in both original image recognition results and enhancement results could be too many non-gene entities from pathway figures, such as protein names, enzyme names, etc. We only focus on the gene names; however, the image recognition results contain all the text entries inside a pathway figure, bringing many non-gene entities.

From Table 1, the enhancement results are better than the image recognition results alone for all four evaluation methods. Among them, the fuzzy-match method with manual curations (i.e., using ground truth as the gene dictionary) leads to the highest performance (as our upper bound), but manual curations are not always available in real applications. Hence, the fuzzy match with PubTator Central retrieved and exHUGO retrieved gene annotations may be useful as both perform better than the whole article. Although the result based on PubTator retrieved has a better recall score, the exHUGO dictionary achieves higher precision. Also, the higher F1 score indicates that exHUGO retrieved results provide better trade-off for gene enhancements.

In addition, based on our experiments, we noticed that the exHUGO dictionary contains many gene names that we did not find in the PubTator Central dictionary, such as MEK1, CED-1, etc. At the same time, different written formats for the same gene could also be considered; for example, NF-KB and NFKB refer to the same gene, where PubTator Central can only identify the latter one. However, in exHUGO, there are also some non-targeted genes included, e.g., CELL, P130Cas, etc.

Table 1. Performance of gene name enhancement

Performance (%)	Image only	Fuzzy match dictionaries			
		Manual curations	Whole article	PubTator central retrieved	exHUGO retrieved
Precision	31.4	40.6	32.8	34.6	66.1
Recall	70.8	91.6	74.0	78.2	68.3
F1	43.5	56.2	45.4	48.0	65.2

4.2 Gene Relation Enhancement Results

The number of annotated gene relations in 45 pathway figures ranges from 1 to 20. Based on our comparison results, the image recognition performance is improved by some of the relation extraction methods. Like gene name enhancements, we listed the average enhancement precisions, recalls, and F1 scores for all three levels, together with the corrected image recognition results, in Table 2.

The "Corrected image only" column indicates that the gene relation results directly from image recognitions are "corrected" by manual curations (ground truth). We filtered out all the non-gene entries in the image recognition results of gene names, leaving only the gene name entries when counting the total gene relations in the pathway figures. TheWeed to do this step is becauseI the other results in Table 2 assume there are only gene name co-occurrences counted, where no other non-gene entries would be counted as gene co-occurrence – which makes the comparisons among different methods fairer. The other columns in Table 2 are the results based on different levels within different ranges of sentences, respectively, inferred from the column name itself.

Table 2. Performance of gene relation enhancement

Performance (%)	Corrected image only	Caption in 1 sentence	Caption in 7 sentences	Paragraph in 1 sentence	Paragraph in 7 sentences	Full text in 1 sentence	Full text in 7 sentences
Precision	47.90	13.88	12.5	15.22	14.76	17.79	16.77
Recall	48.12	33.53	36.08	37.97	40.12	56.81	61.84
F1	48.01	19.63	18.62	21.73	21.58	27.09	26.39

First, from Table 2, although the overall performance for both caption-level and paragraph-level does not increase, the recall scores of both full-text-level in one sentence and in seven sentences are higher than the relations directly extracted from the corrected image only. In the full text of an article, it is likely that more gene co-occurrences in ground truth are detected. For example, (ATR, H2AX), this gene co-occurrence has appeared in both ground truth and the same sentence of full-text-level, but it does not occur in the detected relations from an image. Further, compared to recall scores of the same sentence with the neighboring sentence within three different levels, the recall scores of neighboring sentences are always higher than the same sentence, which implies that the more the sentences involved, the more correct gene co-occurrences could potentially be detected.

Also, for both the same sentence and neighboring sentences, the precision scores are rather lower than the detected relations from the image. The main reason could be that the extra noise is involved in the data while introducing the auxiliary information on both three different levels. For example, in one of the articles containing a certain pathway figure, there are five gene relations in total in the ground truth, the original image recognition only has two corrected gene relations. Whereas the gene co-occurrences detected in the same sentence contain 24 gene co-occurrences, which perfectly match all 5 gene relations in the ground truth, but with 19 irrelevant gene co-occurrences.

Moreover, when comparing within the same level, the precision scores are consistently higher with neighboring sentences than with one sentence only, while the recall scores are completely the opposite. That is because more gene co-occurrences are detected after expanding the "search range", which also involves more "noise genes" simultaneously. For example, for one specific article, there are 7 gene co-occurrences in total within one sentence, but only 1 gene co-occurrence can match the gene relations in the ground truth. After searching gene co-occurrences based on seven sentences, the gene co-occurrences become 9 in total, adding 2 more unmatched gene co-occurrences. As a result, there is still one gene co-occurrence matching the gene relation in the ground truth. Obviously, the 2 more gene co-occurrences can be defined as irrelevant noise.

When comparing the results across the different levels, the recall scores are always increasing, no matter whether gene co-occurrences are detected within 1 sentence or 7 sentences. It indicates that amplifying the "search range" can still match more gene co-occurrences.

Compared to the corrected image recognition results (corrected image only), the precision scores of all gene co-occurrence methods are all decreasing, while only recall scores at full-text level are increasing. The reasons are the following:

1. Some caption-level and paragraph-level sentences do not contain any gene co-occurrences; hence, their gene co-occurrences would be null.
2. Some gene names are different in text and figures. For example, (PKA, GSK3B) is a gene relation in one pathway figure, but GSK3B turns to GSK-3beta in the text. Obviously, GSK3B and GSK-3beta refer to the same gene, but the different expressions hinder the gene co-occurrence detection in the text.

4.3 Use Cases

Since we provided two different enhancements, so the use cases are separated as gene name enhancements and gene relation enhancements. The original figure of Fig. 4 is from [12] and Fig. 5 is from [13].

Gene Name Enhancements. The fuzzy match method with different gene dictionaries can help discover more gene names in the pathway figures – one example is shown in Fig. 4. The blue rectangles in the figure are the gene names found by image recognitions and the red rectangles are the newly discovered ones after applying the fuzzy match method with the PubTator Central [4] retrieved gene names.

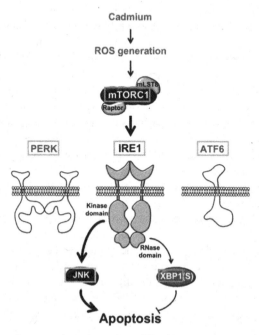

Fig. 4. The use case of gene name enhancement. The blue rectangles indicate the gene names recognized by OCR, where the red rectangles indicate the gene names after applying our pipeline. (Color figure online)

Gene Relation Enhancements. Similar to gene name enhancement, text mining methods can also help discover more gene relations in pathway figures, as illustrated in Fig. 5. The blue rectangles in the figure are gene relations detected by image recognition. The red rectangles are the newly-added one extracted from the text by counting gene co-occurrences in the same sentence.

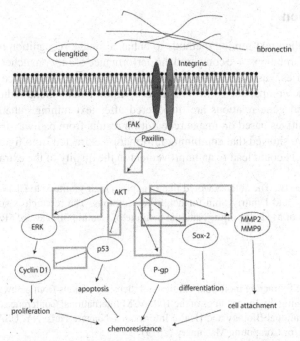

Fig. 5. The use case of gene relation enhancement. The blue rectangles indicate the gene relations recognized by OCR, where the red rectangles indicate the gene relations after applying our pipeline. (Color figure online)

5 Discussion

This work demonstrated that text mining methods could improve the image recognition results of gene names and gene relations. Our current methods for gene co-occurrence detections still exist many problems. For instance, when we identified the sentences for caption-level, paragraph-level, and full-text-level, we utilized the XML file, which might lose some sentences due to the format problem. In the future, we might need to explore other methods to obtain the corresponding sentences.

At the same time, for gene relation enhancements, we will also try to assign weights to different levels of text regions, namely caption-level, paragraph-level and full-text-level. Based on that approach, we should be able to get better performance. Moreover, we plan to extend the exHUGO gene dictionary continuously. Based on a bigger gene dictionary, it could help us identify the same gene name with different expressions in the text, which leads to higher precision and recall scores.

Further work is also ongoing to manually curate more pathway figures to verify that the text mining enhancements of our scientific literature processing framework are generalizable. For example, our next target includes pathway figures from KEGG since the XML ground truth file is already available for each figure. Also, we only utilized existing off-the-shelf text mining tools for both gene name and gene relation extractions, but we plan to explore building our model for this task.

6 Conclusion

This paper applied text mining methods to enhance image recognition results of gene names and gene relations. We compared the performance of fuzzy matches with different "gene dictionaries," as well as the performance of gene co-occurrence gene relation extractions. The comparisons with image recognition results alone indicate that more gene names and gene relations are discovered after text mining enhancements with lower false positives based on image recognition results from pathway figures. Finally, in this paper, we showed that combining information extracted from figures and text of a scientific article could lead to an improvement in the quality of the extracted pathway.

Acknowledgements. The authors would like to express their gratitude to Clement Essien, Drs. Richard Hammer and Dmitriy Shin for helpful discussions. The research is supported by the National Library of Medicine of the National Institute of Health (NIH) award 5R01LM013392.

References

1. He, F., et al.: Extracting molecular entities and their interactions from pathway figures based on deep learning. In: Proceedings of the 10th ACM International Conference on Bioinformatics, Computational Biology and Health Informatics, Niagara Falls, NY, USA, pp. 397–404. Association for Computing Machinery (2019)
2. Hanspers, K., et al.: Pathway information extracted from 25 years of pathway figures. Genome Biol. **21**(1), 273 (2020)
3. Kanehisa, M., et al.: KEGG: new perspectives on genomes, pathways, diseases and drugs. Nucleic Acids Res. **45**(D1), D353–D361 (2016)
4. Wei, C.-H., et al.: PubTator central: automated concept annotation for biomedical full text articles. Nucleic Acids Res. **47**(W1), W587–W593 (2019)
5. Lee, J., et al.: BioBERT: a pre-trained biomedical language representation model for biomedical text mining. Bioinformatics **36**(4), 1234–1240 (2019)
6. Kim, D., et al.: A neural named entity recognition and multi-type normalization tool for biomedical text mining. IEEE Access **7**, 73729–73740 (2019)
7. Kim, M., Baek, S.H., Song, M.: Relation extraction for biological pathway construction using node2vec. BMC Bioinform. **19**(8), 206 (2018)
8. Zhou, J., Fu, B.-Q.: The research on gene-disease association based on text-mining of PubMed. BMC Bioinform. **19**(1), 37 (2018)
9. Braschi, B., et al.: Genenames.org: the HGNC and VGNC resources in 2019. Nucleic Acids Res. **47**(D1), D786–D792 (2018)
10. Tafti, A.P., Baghaie, A., Assefi, M., Arabnia, H.R., Yu, Z., Peissig, P.: OCR as a service: an experimental evaluation of Google Docs OCR, Tesseract, ABBYY FineReader, and Transym. In: Bebis, G., et al. (eds.) ISVC 2016. LNCS, vol. 10072, pp. 735–746. Springer, Cham (2016). https://doi.org/10.1007/978-3-319-50835-1_66
11. Levenshtein, V.I.: Binary codes capable of correcting deletions, insertions, and reversals. Dokl. Akad. Nauk SSSR **10**, 707–710 (1965)
12. Kato, H., Katoh, R., Kitamura, M.: Dual regulation of cadmium-induced apoptosis by mTORC1 through selective induction of IRE1 branches in unfolded protein response. PLoS ONE **8**(5), e64344–e64344 (2013)
13. Yu, Q., et al.: Fibronectin promotes the malignancy of glioma stem-like cells via modulation of cell adhesion, differentiation, proliferation and chemoresistance. Front. Mol. Neurosci. **11**, 130 (2018)

Sentence Classification to Detect Tables for Helping Extraction of Regulatory Interactions in Bacteria

Dante Sepúlveda[ID], Joel Rodríguez-Herrera[ID], Alfredo Varela-Vega[ID],
Axel Zagal Norman[ID], and Carlos-Francisco Méndez-Cruz[✉][ID]

Centro de Ciencias Genómicas, Universidad Nacional Autónoma de México,
Avenida Universidad s/n, 62210 Cuernavaca Mor., Mexico
{dtorres,joelrh,avarela,azagal}@lcg.unam.mx, cmendezc@ccg.unam.mx
https://www.ccg.unam.mx/en/computational-genomics/

Abstract. The biomedical knowledge about transcriptional regulation in bacteria is rapidly published in scientific articles, so keeping biological databases up to date by manual curation is rather than impossible. Despite the efforts in biomedical text mining, there are still challenges in extracting regulatory interactions (RIs) between transcription factors and genes from text documents. One of them is produced by text extraction from PDF files. We have observed that the extraction of RIs from text lines that comes from tables of the original PDF article produces false positives. Here, we address the problem of automatically separating this text lines from those that are regular sentences by using automatic classification. Our best model was a Support Vector Classifier trained with n-grams of characters of tags of parts of speech, numbers, symbols, punctuation, brackets, and hyphens. Despite a significant imbalanced data, our classifier archived a positive class F1-score of 0.87. Our best classifier will be coupled eventually to a preprocessing pipeline for the automatic generation of transcriptional regulatory networks of bacteria by discarding text lines that comes from tables of the original PDF.

Keywords: Information extraction · Transcriptional regulation · Biomedical text mining · Supervised learning · Regulatory interaction · Machine learning

1 Introduction

A transcriptional regulatory network (TRN) gives a global view of the regulatory mechanisms for bacteria to survive under different environmental conditions. A TRN is formed by several regulatory interactions (RIs) between transcription factors (TFs) and regulated genes. TFs affect the initiation of transcriptional regulation by facilitating (activation) or inhibiting (repression) the gene transcription, so a regulatory interaction is formed by a TF, a gene, and an

D. Chicco et al. (Eds.): CIBB 2021, LNBI 13483, pp. 143–157, 2022.
https://doi.org/10.1007/978-3-031-20837-9_12

effect (Table 1). These TRNs are published in biological databases as a valuable resource for experimental and bioinformatic researchers; however, many bacteria still lack TRNs, and most of the existing ones are incomplete [11].

Table 1. Examples regulatory interactions from the *E. coli* TRN published in RegulonDB [28].

TF (regulator)	Gene	Effect of the regulator
AraC	araA	Activator
AraC	araB	Repressor
AraC	araC	Activator

With the purpose of building a TRN using a collection of 3,200 articles of *Salmonella enterica* serovar Typhimurium (*Salmonella*), which is one of the primary pathogens that infect both humans and animals worldwide [12], we have worked on developing a text mining approach to extract RIs from literature of this bacterium. Our linguistic approach follows the concept of *thematic role*, which refers to the various roles that a nominal phrase plays regarding the action described by a verb; in the case of a RI, the TF is the *agent* and the gene is the *patient*. Using simple rules and triplets extracted with Open Information Extraction [2], we have obtained promising results (manuscript in preparation).

In our current pipeline for RI extraction, an article collection is preprocessed by extracting text from the PDF files using an in-house developed text extractor tool. Afterward, traditional NLP preprocessing steps are performed: sentence split, tokenization, lemmatization, and part of speech tagging, using modules from Stanza NLP library [27]. Then, NLP preprocessing output is processed using rules and OpenIE to extract the final RIs and construct a TRN. Using our approach, we are able to extract regulatory interactions from sentences such as:

1. *It is known that prgH is under the regulation of many global regulators, such as HilA, InvF, PhoP, and SirA* [31].
2. *The PhoP-activated mgtA, mgtC, and pagM genes are required for motility on 0.3 % agarose low Mg2+media* [24].

Analyzing the RIs extracted by our approach, we detected false positives produced by the following situations: 1) Extracted RIs referring to another bacterium different from the bacterium of interest. For example, in the literature of *Salmonella*, we can often find RIs of *E. coli*, as this is a model organism commonly used for comparison, 2) RIs extracted from sentences in which the RI is not verified or implies a supposition, for example, *OmpR appears to function as an activator of ssrA* [13], and 3) RIs extracted from text lines that come from tables of the PDF article after text extraction using our in-house PDF text extractor. The last problem is the one that we address here.

Our in-house PDF text extractor extracted tables from the original article as raw text; then, our NLP preprocessing steps transformed it into text lines. For example, in the first row of Table 2, we show part of the original table obtained from [33]. In the second row of Table 2, we display the extracted line from the PDF file, illustrating that the original order of the data was lost. We obtained the same output using some free available libraries to extract text from PDF files.

Table 2. Example of a table before an after text extraction from PDF file with our in-house developed tool and NLP pre-processing steps. Original table adapted from [33]

Original table	Flagella 1 phase-2 flagellin fljB 52.54 4.75 306.0
	2 phase-1 flagellin fliC 51.34 4.76 2.2
Text line	Flagella phase-2 flagellin phase-1 flagellin 306.0 2.2 2.1 2.3 2.2...

In addition to the text file, our pipeline delivers a semi-structured output (JSON file) used to display articles in a curation Web system [10]. The JSON file includes article sections and metadata that allow us to show a parallel visualization of the PDF article and the corresponding rendering version for curation (Fig. 1. Moreover, the approach to extract RIs has to correctly associate the extracted RIs with the sentences in the JSON file to be displayed in the platform, as extracted RIs are uploaded to the platform for the curator to validate them.

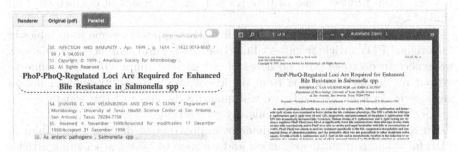

Fig. 1. Parallel visualization of a PDF article and corresponding rendering version in the curation Web system.

In the past, different approaches have been adopted to deal with tables from articles in PDF or HTML format. However, these approaches were oriented towards finding graphic patterns of tables, highlighting the importance of the visual layout or the metadata. Also, previous works have addressed the problem with Conditional Random Fields (CRFs) for predicting patterns and structures of tables through graphical modeling [26]. In another work, a new box-cutting graphical recognition model was implemented to detect adjacent and connected lines that delimit a table crawled from digital libraries [20]. There is a whole field of Web table processing established [35]. A remarkable example is the classification of tables from academic papers, with particular appeal for topics such as biological interactions, chemical patents, or similar fields, which require more extensive and more complex tables than those found on the web [34].

Despite these efforts to deal with tables in PDF format, as a first approach to decrease false-positive RIs, without the need to make significant changes in our in-house text extractor, we propose to classify text lines into two categories: 1) NON-TABLE, if they appear to be regular sentences, and 2) TABLE if they seem to come from a table in the original article. Once all TABLE instances are categorized, they will be discarded in the RI extraction (Fig. 2).

We tested three classical text classifiers: Support Vector Classifier (SVC), Random Forest (RF), and Stochastic Gradient Descent Classifier (SGDC), to find the best model to automatically detect text lines of TABLE category in the extracted text from the PDF files. We employed a manually categorized data set of 150 articles from the literature on transcriptional regulation of *Salmonella*. One of the major challenges in our task was the extensive imbalance in our data set, which consisted of 410 instances of TABLE category (positive class) and 45,406 instances of NON-TABLE category (negative class). We addressed this problem by testing the following strategies: random oversampling, automatic weighting classes inversely proportional to class frequencies, and training based on positive class F1-score. To represent instances of both categories, we used tags of: Part-Of-Speech, numbers, symbols, hypens, brackets, and punctuation. In addition, we examined sentence length as a feature for training. A feature selection strategy was also considered. The best classifier was a SVC with an F-score of the positive class of 0.87 trained using n-grams of characters of all kind of tags.

This work will help us separate text lines that appear to come from tables in the original PDF article from text lines that are natural sentences. In this way, we expect to decrease false positives in our current (unpublished) pipeline to extract regulatory interaction from the literature on transcriptional regulation in bacteria. In a future study, we will quantify the effect of this separation on the performance of extracting regulatory interactions.

In future work, we will test different approaches to dealing with tables in PDF files. The automatic extraction and curation of tables in scientific articles remain a need that increases proportionally with the number of new publications on transcriptional regulation. We predict this behavior based on the ability of tables to store dense and multidimensional information, something useful for the presentation of complex interaction networks. Another characteristic of tables is that they are displayed as easily understandable structures to humans but could be algorithmically complex to analyze because of the variety of layouts.

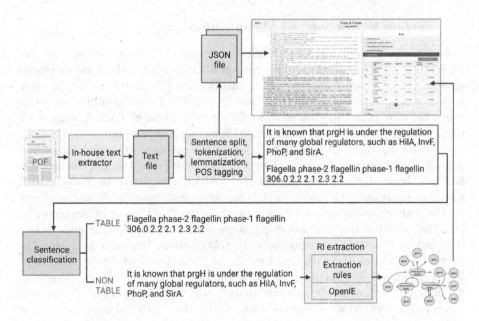

Fig. 2. RI extractor pipeline. Created with BioRender.com

2 Materials and Methods

2.1 Data Set

We obtained a list of 3,200 PubMed IDs (PMIDs) using a PubMed query based on keywords of transcriptional regulation in *Salmonella*. This list was used to download the PDF files. We extracted text files from these PDF articles using our in-house tools. From the collection of text files, we selected a set of 150 files by searching the word *table*. We manually checked the text lines of the 150 articles where the search word *table* was found to confirm that these lines came from tables in the original PDF article, avoiding references to tables in the supplementary material. These text lines, which come from tables, were considered the positive class, which we categorized as TABLE (see Table 2). The remaining text lines were considered the negative class and labeled NON-TABLE. Therefore, the data set comprised 410 instances of the positive class and 45,406 of the negative class. As we mentioned, the imbalanced data problem was the main challenge of this work since the number of examples of the positive class was much smaller than those of the negative class. The data set was split into 80% for training and 20% for testing the model.

2.2 Feature Extraction and Vectorization

In traditional machine learning, the representation of training instances by useful features is essential to get the best predictive model; thus, we employed some commonly used features in text classification [23]. After inspecting some instances of both categories (TABLE and NON-TABLE), we identified some features that were primarily present in text lines categorized as TABLE, suggesting that they may be useful for classification. Some of those were numbers, symbols, parenthesis, brackets, punctuation (.,;), and the presence of several nouns and prepositions. However, many of these features were present in both categories; for example, small text lines categorized as TABLE were similar to NON-TABLE text lines expressing references, as they also contained parenthesis and numbers.

For feature extraction, we represent instances (text lines) as sequences of tags that resulted from preprocessing steps with Stanza [27]. The annotations returned by Stanza were: sentence split, tokenization, lemmatization, and the Part-Of-Speech tagging with tags used in the Penn Treebank Project [22]. This POS tagging includes linguistic elements (nouns, verbs, prepositions) as well as cardinal numbers, symbols, and punctuation [14]. In Table 3 an instance of each category represented with tags is shown. Notice that both instances share some tags.

Table 3. Examples of the final representation of input data.

NON-TABLE	NN IN NN VBZ DT JJ NN IN DT NN -LRB- NNP FW FW, CD -RRB-, CC WRB DT NN NN HYPH NNS NN NN IN NN CC NN VBZ RB
TABLE	NNP NN HYPH NN NNS IN DT NN HYPH NN NN VBN IN NNP NNP -LRB- NNP -RRB- IN DT JJ HYPH NN NN -LRB- NN -RRB-, NN NN NN -LRB- NNP -RRB- ...

Abbreviations: coordinating conjunction (CC), cardinal digit (CD), determiner (DT), adjective (JJ), foreign word (FW), preposition(IN), hyphen (HYPH), singular noun (NN), plural noun (NNS), proper noun (NNP), adverb (RB), symbol (SYM), predeterminer (PDT), left square bracket (-LRB-), right square bracket (-RRB-), verb (VB), verb in past participle (VBN), verb gerund (VBG), verb in 3rd person singular present (VBZ), wh- pronoun (WP), wh- adverb (WRB).

We tested four representations of instances based on tags: 1) symbol tags, 2) number tag, 3) number and symbol tags, and 4) all tags (Table 4). We also considered sequential combinations of tags (n-grams) from these representations to train the classifiers, hypothesizing that some combination of tag patterns may help separate between instances of each category, for example, sequences of numbers or combinations of nouns and numbers. We trained with individual tags (n-grams = 1), two sequential tags (n-grams = 2), three sequential tags (n-grams = 3), one and two tags (n-grams = 1,2), two and three tags (n-grams = 2,3), and one, two and three tags (n-grams = 1,3). An example of tag n-grams from all tags representation is shown in Table 4. Since another observation was that text lines categorized as TABLE seemed to be longer than those categorized as

NON-TABLE, we tested the length of the instance (total number of characters) as a training feature.

Table 4. Different representations of instance features.

All tags	NN IN NN VBZ DT JJ NN IN DT NN ... -LRB- NNP FW
Tag n-grams	(NN IN) (NN VBZ) (DT JJ) (NN IN) (DT NN) ... (NNP FW)
Symbol tags	LRB RRB LCB RCB SYM HYPH NFP, . :
Number and symbol tags	CD LRB RRB LCB RCB SYM HYPH, . :
Characters	N N I N N N V B Z D T J J N N I N ... N N P F W
Character n-grams	(N N _ I) (I N _ N) (N N _ V) (D T _ J) ... (P _ F W)
Instance length	49

Abbreviations: left curly brackets (LCB), right curly brackets (RCB)

We also included using sequences of characters of tags (character n-grams) as another way of representing instances. The main idea behind this representation is to reduce features, which also helps to save computational costs. This representation of character-level k-grams has been used previously in text classification, such as language identification [17]. We represented instances with two and three characters (n-grams = 2,3), two, three and four characters (n-grams = 2,4), two, three, four and five characters (n-grams = 2,5), three and four characters (n-grams = 3,4), and three, four and five characters (n-grams = 3,5). An example of character n-grams is shown in Table 4. Although this representation seems unusual, we obtained the best results training with this representation.

All instances were vectorized to a term frequency-inverse document frequency matrix (TF-IDF) of size $(n - instances, m - features)$, in which every row was the numerical vector representation of a single instance. The TF-IDF has been proposed in information retrieval [29] to give more importance to features that best describe an instance, and it is widely used in NLP problems [32]. The weight is very low for features (i.e., tags, n-grams of tags, n-grams of characters) that occur in most instances, while it is higher for those which only occur in some of them. This matrix and an array of length $n - instances$ containing the class category associated with each instance were provided to the learning algorithm. All the classifiers were trained with all the vectorized representations mentioned above.

It is well known that vector representations of text data are highly dimensional [32]. Thus, we tested a dimensionality reduction strategy by choosing a percentile of 70% or 90% of the features from the TF-IDF matrix. Dimensionality reduction has benefits in memory, computation, complexity, variance reduction, and elimination of noise and outliers [1]. This strategy removes features until reaching a user-defined percentage of the highest-scoring features based on a chosen metric. For scoring features, we employed the f_classif, which computes the ANOVA F-value between classes and features. By doing this, we manage to train with only the most relevant features. We utilized the *SelectPercentile* method of scikit-learn (https://scikit-learn.org/stable/) [25] to implement this strategy.

2.3 Supervised Learning

We compared the performance of three classical classification algorithms that have previously proven effective in text classification [16]: Random Forest Classifier, Stochastic Gradient Descent Classifier, and Support Vector Classifier. We opted for tools available in Scikit-Learn for text preprocessing, feature extraction, and classification since this library offers one of the best-integrated ecosystems and flexible pipelines for machine learning [30].

Decision trees are simple structures used to solve complex problems. The Random Forest Classifier (RFC) uses several decision trees to fit sub-samples of the data provided [7]. The maximum size of each sub-sample can be modified via the parameter n_estimators. The run time can be pretty short for binary classification compared to multi-class classification since there are significantly few decisions per node.

The Stochastic Gradient Descent Classifier (SGDC) focuses on the gradient of loss estimated in each sample, which updates the model in real-time to alter the weight given to each classified sample using a modifiable learning rate [6]. This method lets us alter the learning rate (learning_rate), the regularisation term (penalty), and the loss function (loss) used by the model, making this a highly adjustable gradient.

The Support Vector Classifier (SVC) finds the best decision hyperplane that separates vectors representing instances of the two classes [9]. SVC accomplishes this by finding the pair of instances that is the hardest to distinguish (support vectors) and using them to build the decision function, which is unique for a training set [5]. We used a Radial Basis Function (RBF) as a kernel for the SVC, which uses the hyperparameters C and Gamma to calculate how related two data points are and give them a class based on that closeness.

The imbalance between the number of positive and negative instances is a huge challenge in machine learning [18]; however, there are no theory-based guidelines to pick the best strategy to deal with this problem. We tried to address our imbalanced data problem by testing three strategies: 1) the random oversampling of the positive instances, 2) automatically adjusting weights of classes inversely proportional to the frequency of the instances of each class, and 3) training classifiers based on positive class performance score.

To balance our data, we performed a random oversampling. We generate new instances by randomly sampling with the replacement of the currently available instances in the positive class. We used the Python library imbalanced-learn to implement this strategy (https://imbalanced-learn.org) [19]. The random oversampling was performed excursively on training data, and the model was evaluated with imbalanced test data as it is suggested in literature [21].

The second strategy we employed to address the imbalanced problem was the weighting of classes inversely proportional to the frequency of the instances of each category. This step modifies the training of the algorithms by penalizing the classification errors of the minority class. This strategy is already implemented in the models of the Scikit-Learn [25].

In addition, to deal with this imbalanced data problem, we trained the classifiers by evaluating the F-1 score value instead of the accuracy, which is unsuitable for evaluating classifiers when the imbalanced problem is present [3]. We trained the classifiers by optimizing only the F-1 score of the positive class (TABLE) as a way to search for an improvement in the classification performance of this category. We also trained classifiers using a weighted F1-score metric, calculated for each category and then weighted by the number of instances of each class. This allowed us to alter the F1-score to account for imbalanced data (the F1-score is not between precision and recall).

All training runs were performed with the stratified k-fold cross-validation strategy, with $k = 5$, and searching of hyperparameters: C and gamma for the SVC; alpha, penalty, and l1 ratio for the SGDC; and estimators, bootstrapping, and criterion for the RFC. A randomized search over parameter values guided the optimization of hyperparameters by sampling from a distribution over possible parameter values. This strategy has the benefit of adding parameters without having a decrement in efficiency [4]. We took advantage of the *RandomizedSearchCV* object implemented in Scikit-Learn [25] for the optimization of hyperparameter over 100 iterations. In Table 5, we show the range of values utilized for optimization for each hyperparameter.

Table 5. Parameter settings for optimisation of hyperparameters (100 iterations)

Algorithm	Hyperparameter	Values
SVC	C	1–50, step 0.5
	Gamma	0–1, step 0.05
SGDC	Alpha	10^x: x from 0 to −7
	l1,l2	Ratio: 0.15, 0.25, 0.5, 0.75
RFC	Estimators	100, 150, 200, 300
	Bootstrap	True, false
	Criterion	Gini,entropy

3 Results

3.1 Best Model

In Table 6, we summarize all the features used for instance representation, the parameters employed for training classifiers, and the combination of them that obtained the best performance (third column). To understand how every parameter influences one another and to measure their impact on the classifier's performance, we tested all possible combinations, giving us a total of 397 cases (runs).

Table 6. Summary of features used for instance representation and parameters employed for training classifiers. We include the combination that obtained the best performance in third column.

Features and Parameters	Tested options	Best option
Features	All tags, Only symbols, Only numbers, Symbols + numbers, instance length	All tags
Vectorization	Tag, Character	Character
N-gram range	Tag:(1,1),(2,2),(3,3),(1,2),(2,3),(1,3)	Character (2–5)
	Character:(2,3),(2,4),(2,5),(3,4),(3,5)	
Feature Selection	None,70%, 90%	None
Oversampling	True, False	False
Scorer	Weighted, Only Positive Class	Weighted

Oversampling of the positive class, the use of instance length as a feature, and only training with number tags or symbol tags, did not significantly increase the model performance. On the contrary, the weighting of classes inversely proportional to the frequency of the instances of each category (Scorer Weighted) to handle the imbalanced data and using character n-grams was an essential step for enhancing model behavior since the highest 24 F1-scores on the test data were archived by models using this configuration (Table 7). Surprisingly, neither training with tag n-grams nor single tags led to anything close to outstanding; the highest positive class F1-score on test data obtained with this strategy stood in 24th place.

Table 7. Best positive class F1-scores obtained in test data.

Rank	Run ID	Recall	Precision	F1-Score	MCC	Feature	N-gram range
1	Run 289 (SVC)	0.88	0.85	0.87	0.87	Character	(2,5)
2	Run 291 (SVC)	0.88	0.84	0.86	0.86	Character	(2,5)
3	Run 390 (SVC)	0.87	0.84	0.86	0.85	Character	(2,5)
4	Run 363 (SGDC)	0.87	0.85	0.86	0.86	Character	(3,4)
5	Run 328 (SVC)	0.82	0.88	0.85	0.85	Character	(3,5)
...	...	...	...	...	...	...	...
24	Run 109 (SVC)	0.86	0.80	0.82	0.82	Tag	(1,2)

The best classifier was obtained with an SVC using from 2 to 5 character n-grams and without feature selection. Best hyperparameters were $C = 23$ and $gamma = 0.41$. Despite the imbalanced data problem, this classifier obtained an F-score for the positive class of 0.87 on the test data. To further back up these results, we measured the Mattews Correlation Coefficient (MCC) [8] for predictions on the test data and obtained an MCC score of 0.87. This metric is a reliable indicator of the quality of a binary classification with a highly imbalanced data set because it measures the correlation between the true and the predicted values. Thus, the obtained MCC score is consistent with the model performance observed with the F1-score even though positive instances were significantly underrepresented. Data and models are available in https://github.com/laigen-unam/table-in-text.

We also used the Precision-Recall curve and the Receiver Operator Characteristic (ROC) curve for evaluation. These evaluations have proven helpful for counterfactual cases with imbalanced and balanced training data sets, respectively [15]. We evaluated both and plotted the curve showing some metrics (Fig. 3). We observed that the area under (AU) the ROC curve (Fig. 3a) appeared to be optimistic, but it may mask a poor performance caused by the imbalance of our data. However, the AU of the Precision-Recall curve is a more realistic metric of our classification performance and showed that it got acceptable average precision (AP = 0.93) (Fig. 3b). The confusion matrix (Fig. 3c) on the test data for the best classifier highlights our relevant results. Although there was a significant difference in the number of instances between classes, only 9% of the TABLE instances were classified wrongly in the NON-TABLE category.

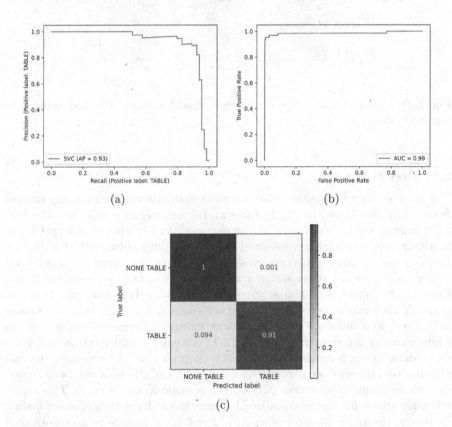

Fig. 3. Results for the best classifier: (a) ROC curve. (b) Precision-Recall curve. (c) Confusion matrix.

As we observed that the best combination of features was characters in the n-gram range of (2,5) (Table 7), we explored increasing the number of iterations to 700 for optimizing hyperparameters by training the three classifiers. Results showed an increment in positive class F1-score. The SVC obtained a positive class F1-score of 0.90, surpassing the RFC and the SGDC (Fig. 4).

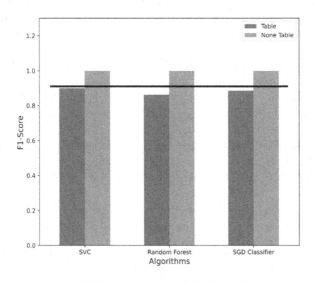

Fig. 4. Best positive class F1-Scores for each algorithm on the test data using (2,5) n-grams of characters.

3.2   Best Features

The features used by the best classifier were character n-grams ranging in sizes from 2 to 5. We show the top features in Table 8 to gain insight into the patterns learned by the best model. As we expected, features like numbers (CD) or punctuation symbols (:) were essential for distinguishing between the two classes. Interestingly, dividing by n-grams of characters instead of n-grams of tags helped identify combinations of sequences and repeated elements, as we can see in features 1–6 in Table 8. Notice that spaces were considered as characters to create n-grams, allowing us to preserve the tag information. By doing this, we manage to capture more information than the one captured by complete tag n-grams while avoiding the computational cost and the exponential increase in feature dimensionality that comes with increasing n-gram size. For example, line #5 captures sequences of two consecutive numbers (CD CD) and a number followed by a coordinating conjunction (CD CC) in a single feature (cd c). This representation allows for more generalized features and reduces their dimensionality. Moreover, the space at the beginning of the (cd c) feature retains positional information, suggesting this sequence of tags does not appear at the start of the text line.

Table 8. Top 1% best features using n-grams of size (2,5). Note that features 37–39 contain the information of 3 tags while being smaller that complete tag n-gram of size (3–3) displaying the same sequence of elements.

Rank	Feature	N-gram size	Tags feature
#1	(. n)	4	2
#2	(. nn)	5	2
#3	(: :)	4	2
#4	(: :)	5	2
#5	(cd c)	5	2
...	...	...	...
#37	(d : :)	5	3
#38	(n . n)	5	3
#39	(n : :)	5	3

4 Conclusion

By training a classifier, we addressed the problem of automatically separating text lines that appear sentences from those that came from tables in PDF files. The best model archived an F-score of the positive class of 0.87 on the test data, improvable to 0.90. It was obtained by a Support Vector Classifier trained with n-grams of characters of parts of speech tags, numbers, symbols, punctuation, brackets, and hyphens. This classifier will be coupled to our pipeline to extract regulatory interactions from the literature on transcriptional regulation in bacteria to evaluate its effect on reducing false positives. This work may have an impact on extracting TRNs for bacteria without an available curated database.

Acknowledgments. This work was supported by UNAM-PAPIIT IA203420 and the Universidad Nacional Autónoma de México (UNAM). We acknowledge Víctor del Moral Chávez and Alfredo José Hernández Álvarez for computational support.

References

1. Alpaydin, E.: Introduction to Machine Learning. MIT Press, Cambridge (2020)
2. Angeli, G., Johnson Premkumar, M.J., Manning, C.D.: Leveraging linguistic structure for open domain information extraction. In: Proceedings of the 53rd Annual Meeting of the Association for Computational Linguistics and the 7th International Joint Conference on Natural Language Processing, vol. 1, pp. 344–354. Association for Computational Linguistics, Beijing (2015). https://doi.org/10.3115/v1/P15-1034
3. Bekkar, M., Djemaa, H.K., Alitouche, T.A.: Evaluation measures for models assessment over imbalanced data sets. J. Inf. Eng. Appl. **3**(10), 27–39 (2013)
4. Bergstra, J., Bengio, Y.: Random search for hyper-parameter optimization. J. Mach. Learn. Res. **13**(2), 281–305 (2012)

5. Bishop, C.M.: Pattern Recognition and Machine Learning, p. 738. Springer, NY (2006)

6. Bottou, L.: Stochastic gradient descent tricks. In: Montavon, G., Orr, G.B., Müller, K.-R. (eds.) Neural Networks: Tricks of the Trade. LNCS, vol. 7700, pp. 421–436. Springer, Heidelberg (2012). https://doi.org/10.1007/978-3-642-35289-8_25

7. Breiman, L.: Random forests. Mach. Learn. **45**(1), 5–32 (2001). https://doi.org/10.1023/A:1010933404324

8. Chicco, D., Jurman, G.: The advantages of the Matthews correlation coefficient (MCC) over F1 score and accuracy in binary classification evaluation. BMC Genom. **21**(1), 1–13 (2020). https://doi.org/10.1186/s12864-019-6413-7

9. Cortes, C., Vapnik, V.: Support-vector networks. Mach. Learn. **20**(3), 273–297 (1995). https://doi.org/10.1007/BF00994018

10. Díaz-Rodríguez, M., et al.: Lisen&Curate: a platform to facilitate gathering textual evidence for curation of regulation of transcription initiation in bacteria. Biochim. Biophys. Acta, Gene Regul. Mech. **1864**(11), 194753 (2021). https://doi.org/10.1016/j.bbagrm.2021.194753

11. Escorcia-Rodríguez, J.M., Tauch, A., Freyre-González, J.A.: Abasy Atlas v2.2: the most comprehensive and up-to-date inventory of meta-curated, historical, bacterial regulatory networks, their completeness and system-level characterization. Comput. Struct. Biotechnol. J. **18**, 1228–1237 (2020). https://doi.org/10.1016/j.csbj.2020.05.015

12. Fàbrega, A., Vila, J.: Salmonella enterica serovar Typhimurium skills to succeed in the host: virulence and regulation. Clin. Microbiol. Rev. **26**(2), 308–341 (2013)

13. Feng, X., Oropeza, R., Kenney, L.J.: Dual regulation by phospho-OmpR of ssrA/B gene expression in Salmonella pathogenicity island 2. Mol. Microbiol. **48**(4), 1131–1143 (2003). https://doi.org/10.1046/j.1365-2958.2003.03502.x

14. Ferrario, A., Nagelin, M.: The art of natural language processing: classical, modern and contemporary approaches to text document classification. Modern and Contemporary Approaches to Text Document Classification (March 1, 2020) (2020)

15. Jeni, L., Cohn, J., De la Torre, F.: Facing imbalanced data – recommendations for the use of performance metrics. In: Proceedings - 2013 Humaine Association Conference on Affective Computing and Intelligent Interaction, ACII 2013, vol. 2013, pp. 245–251 (2013). https://doi.org/10.1109/ACII.2013.47

16. Kadhim, A.I.: Survey on supervised machine learning techniques for automatic text classification. Artif. Intell. Rev. **52**(1), 273–292 (2019). https://doi.org/10.1007/s10462-018-09677-1

17. Konheim, A.G.: Cryptography, a Primer. Wiley, Chichester (1981)

18. Kubat, M., Matwin, S., et al.: Addressing the curse of imbalanced training sets: one-sided selection. In: Icml, vol. 97, p. 179. Citeseer (1997)

19. Lemaître, G., Nogueira, F., Aridas, C.K.: Imbalanced-learn: a Python toolbox to tackle the curse of imbalanced datasets in machine learning. J. Mach. Learn. Res. **18**(17), 1–5 (2017)

20. Liu, Y., Bai, K., Mitra, P., Giles, C.L.: TableSeer: automatic table metadata extraction and searching in digital libraries. In: Proceedings of the 7th ACM/IEEE-CS Joint Conference on Digital Libraries, pp. 91–100 (2007)

21. Lusa, L., et al.: Joint use of over-and under-sampling techniques and cross-validation for the development and assessment of prediction models. BMC Bioinform. **16**(1), 1–10 (2015)

22. Marcus, M.P., Marcinkiewicz, M.A., Santorini, B.: Building a large annotated corpus of English: the Penn treebank. Comput. Linguist. **19**(2), 313–330 (1993)

23. Moschitti, A., Basili, R.: Complex linguistic features for text classification: a comprehensive study. In: McDonald, S., Tait, J. (eds.) ECIR 2004. LNCS, vol. 2997, pp. 181–196. Springer, Heidelberg (2004). https://doi.org/10.1007/978-3-540-24752-4_14

24. Park, S.Y., Pontes, M.H., Groisman, E.A.: Flagella-independent surface motility in Salmonella enterica serovar Typhimurium. Proc. Natl. Acad. Sci. **112**(6), 1850–1855 (2015). https://doi.org/10.1073/pnas.1422938112

25. Pedregosa, F., et al.: Scikit-learn: machine learning in Python. J. Mach. Learn. Res. **12**(85), 2825–2830 (2011)

26. Pinto, D., McCallum, A., Wei, X., Croft, W.B.: Table extraction using conditional random fields. In: Proceedings of the 26th Annual International ACM SIGIR Conference on Research and Development in Informaion Retrieval, pp. 235–242 (2003)

27. Qi, P., Zhang, Y., Zhang, Y., Bolton, J., Manning, C.D.: Stanza: a Python natural language processing toolkit for many human languages. In: Proceedings of the 58th Annual Meeting of the Association for Computational Linguistics: System Demonstrations (2020)

28. RegulonDB: Regulatory network interactions (2022). http://regulondb.ccg.unam.mx/menu/download/datasets/index.jsp. Accessed 19 June 2022

29. Sparck Jones, K.: A statistical interpretation of term specificity and its application in retrieval. J. Doc. **28**(1), 11–21 (1972)

30. Varoquaux, G., Buitinck, L., Louppe, G., Grisel, O., Pedregosa, F., Mueller, A.: Scikit-learn: machine learning without learning the machinery. GetMobile: Mob. Comput. Commun. **19**(1), 29–33 (2015). https://doi.org/10.1145/2786984.2786995

31. Wang, L., et al.: InvS coordinates expression of PrgH and FimZ and is required for invasion of epithelial cells by Salmonella enterica serovar Typhimurium. J. Bacteriol. **199**(13), e00824-16 (2017). https://doi.org/10.1128/JB.00824-16

32. Weiss, S.M., Indurkhya, N., Zhang, T., Damerau, F.: Text Mining: Predictive Methods for Analyzing Unstructured Information. Springer, NY (2010). https://doi.org/10.1007/978-0-387-34555-0

33. Yoon, H., Lim, S., Heu, S., Choi, S., Ryu, S.: Proteome analysis of Salmonella enterica serovar Typhimurium fis mutant. FEMS Microbiol. Lett. **226**(2), 391–396 (2003)

34. Zhai, Z., et al.: ChemTables: a dataset for semantic classification on tables in chemical patents. J. Cheminformatics **13**(1), 97 (2021)

35. Zhang, S., Balog, K.: Web table extraction, retrieval, and augmentation: a survey. ACM Trans. Intell. Syst. Technol. **11**(2), 1–35 (2020). https://doi.org/10.1145/3372117

RF-Isolation: A Novel Representation of Structural Connectivity Networks for Multiple Sclerosis Classification

Antonella Mensi[1]([✉])[ID], Simona Schiavi[1,2][ID], Maria Petracca[3,4],
Nicole Graziano[4], Alessandro Daducci[1], Matilde Inglese[2,4],
and Manuele Bicego[1]

[1] Department of Computer Science, University of Verona, Verona, Italy
antonella.mensi@univr.it
[2] Department of Neuroscience, Rehabilitation, Ophthalmology, Genetics, Maternal
and Child Health (DINOGMI), University of Genoa, Genoa, Italy
[3] Department of Human Neurosciences, Sapienza University of Rome, Rome, Italy
[4] Department of Neurology, Icahn School of Medicine at Mount Sinai, New York,
NY, USA

Abstract. Magnetic Resonance Imaging (MRI) is one of the tools used
to identify structural and functional changes caused by multiple sclerosis,
and by processing MR images, connectivity networks can be obtained.
The analysis of structural connectivity networks of multiple sclerosis
patients usually employs network-derived metrics, which are computed
independently for each subject. We propose a novel representation of
connectivity networks that is extracted from a model trained on the
whole multiple sclerosis population: RF-Isolation. RF-Isolation is a vector encoding the disconnection of each region of interest with respect to
all other regions. This feature can be easily captured by isolation-based
outlier detection methods. We therefore reformulate the task as an outlier detection problem and propose a novel approach, called MS-ProxIF,
based on a variant of Isolation Forest, a Random Forest-based outlier
detection system, from which the representation is extracted. We test
the representation via a set of classification experiments, involving 79
subjects, 55 of which suffer from multiple sclerosis. In particular, we
compare favourably to the most used network-derived metrics in multiple sclerosis.

Keywords: Multiple sclerosis · Structural connectivity network ·
Microstructure informed tractography · Proximity isolation forest

1 Introduction

Multiple Sclerosis (MS) is a chronic autoimmune disease of the central nervous
system causing demyelination and neurodegeneration [1]. It is usually diagnosed
and followed up via the analysis of Magnetic Resonance Imaging (MRI), which is
sensitive to demyelination, i.e. lesions affecting white matter tracts and causing
disconnection of grey matter regions.

D. Chicco et al. (Eds.): CIBB 2021, LNBI 13483, pp. 158–169, 2022.
https://doi.org/10.1007/978-3-031-20837-9_13

The aim of this paper is to discern MS patients from healthy subjects, starting from their quantitative connectivity networks estimated with MRI. A connectivity network encodes the connectivity strength between each pair of brain regions of interest (ROIs). An interesting technique, used to compare MS subjects to a healthy cohort, consists of analyzing the distribution of network-derived metrics. These measures capture the importance in terms of connectivity of either the whole network or a subnetwork composed by fewer ROIs. The rationale behind these methodologies is that in connectivity networks, because of the presence of lesions, the connectivity strength between the ROIs involved in a lesion is expected to be lower in an MS subject than in a healthy one. Although the relevance of network-derived measures in the study of MS has been already investigated [11,14,15], there are some issues that would benefit from methodological improvement. In particular, all these measures are subject-wise: they are computed independently for each subject using only the connectivity network of the subject under analysis. In other words, they tend to capture how MS behaves in a specific subject, rather than the global nature of the disease.

In this paper, we propose a novel representation of MS connectivity networks, called *RF-Isolation*. The main characteristics of *RF-Isolation* are: i) it measures the disconnection of each ROI with respect to all other ROIs; ii) the problem is reformulated as an outlier detection task, hence the disconnection of a ROI is represented via an outlierness score, extracted by using a Random Forest (RF)-based outlier detector model; iii) it is disease-wise, i.e. it is extracted for all subjects, MS and healthy, from a model built using the entire MS population.

We assessed the suitability and robustness of the proposed approach via several classification experiments, performed on a cohort of 79 subjects, 55 of which suffer from MS. We also compared to standard network-derived metrics.

2   Materials and Methods

2.1   Study Population

The dataset we employ consists of 79 subjects[1]: 55 suffering from MS and 24 healthy controls. MS subjects had to meet several inclusion criteria: for example, among others [15], their Expanded Disability Status Scale score had to be ≤ 7 and they had to satisfy the diagnostic McDonald criteria. There were also several exclusion criteria, such as: the presence of any major systemic condition, pregnancy, and addiction to drugs/alcohol, among others [15]. Further, they underwent a clinical examination within a week from the MRI scan. All subjects have signed a written informed consent prior to the beginning of the whole study, as the Declaration of Helsinki states. The Institutional Review Board of the Icahn School of Medicine at Mount Sinai approved the protocol.

---

[1] The dataset has been collected at the Mount Sinai Hospital of New York (US) by the group of Matilde Inglese. The dataset is not publicly available.

2.2   MRI Acquisition and Processing

The MRI acquisition protocol was the same for all subjects: a Siemens Skyra 3T scanner (Siemens, Erlangen, Germany) with a 32-channels head coil was used –further technical details can be found in [15]. As to the processing of the brain images, the first part of the pipeline illustrated in Fig. 1 of [14] was followed. Briefly, images were segmented obtaining a cortical parcellation in 85 grey matter ROIs using the Desikan-Killiany atlas [6]. At the same time, the tractography was computed using a probabilistic algorithm. The next crucial step was the application of the COMMIT (Convex Optimization Modeling for Microstructure Informed Tractography) framework [4] to obtain quantitative structural connectivity networks that better reflect the white matter tissue microstructure.

The final connectivity networks, one for each subject, were obtained by combining the COMMIT-weighted tractogram with the segmented grey matter. Each connectivity network is composed by 85 brain regions of interest of grey matter, encoded as nodes; the connectivity strength between each pair of ROIs is encoded as an undirected edge between the involved ROIs. Please note that in the rest of the paper, we will use the terms ROIs and brain regions interchangeably.

2.3   RF-Isolation Extraction

The proposed methodology, *RF-Isolation*, is based on three assumptions:

i *RF-Isolation* is a vector of length equal to the number of established brain regions. Each feature of the vector encodes the degree of disconnection of one ROI with respect to all other ROIs. Indeed, disconnections in the brain are a cornerstone of MS.

ii To encode such characteristic, we map the problem to the outlier detection context. Outliers are objects which do not conform to the rest of the data: they are few and different from the former. We can easily interpret a ROI with a high level of disconnection as an outlier. Therefore, the disconnection level of a ROI can be measured by quantifying its "outlierness".

iii The last assumption is that to extract a meaningful representation for MS, we should build the model using only subjects suffering from MS. Using only the healthy cohort, or both populations, would lead to a model in which the identification of disconnected ROIs is presumably more difficult.

In detail, our approach is based on two steps:

1. Train a model using the *entire MS population*.
2. Given a subject, we can obtain its *RF-Isolation* vector by employing the model from Step 1.

Before thoroughly describing each step, a note must be made on how the adopted model was chosen: our reasoning starts from Isolation Forest (iForest) [8], an RF-based methodology that is also one of the most successful outlier detectors. The aim of iForest is to separate each object from the rest of the data, relying on the principle of isolation. Outliers, due to their nature, are likely to get isolated early in a tree, i.e. they have a high isolation capability. The latter is used to quantify the anomaly score, i.e. outlierness degree, of an object. Unfortunately, iForests work only with vectorial data, whereas our starting point is a connectivity network, which can be seen as a similarity matrix. In detail, the connection strength between two ROIs can be interpreted as a similarity value, i.e. the stronger the connection, the higher the similarity. Thus, we based our approach on Proximity Isolation Forest (ProxIF) [9], an RF and isolation-based outlier detection methodology that works with all types of data for which a proximity measure can be defined. In detail, we extend ProxIF and adapt it for this applicative context: we denote the obtained model as *MS-ProxIF*.

Figure 1 depicts the pipeline of the proposed approach. In Fig. 1(a) we illustrate the building procedure of an MS-ProxIF model, whereas in Fig. 1(b) we use the MS-ProxIF model from Fig. 1(a) to extract the *RF-Isolation* representation. Each step is thoroughly described in the next two subsections.

Step 1. The first step, i.e. the building procedure of MS-ProxIF, is the most complex. A ProxIF is an ensemble learner composed by several randomized decision trees, called Proximity Isolation Trees (ProxITs). Each ProxIT is built using a similarity matrix encoding pairwise similarities, which in our context, corresponds to building an MS-ProxIT using one connectivity network. Actually, differently from ProxIT, each MS-ProxIT can be built using several connectivity networks, each representing a different subject. This extension is crucial, since each MS-ProxIT, rather than capturing subject-wise characteristics, can retrieve disease-specific information. Therefore, an MS-ProxIT is built using a random subset of both ROIs and connectivity networks, each drawn without replacement.

Before describing how to build an MS-ProxIT, we must define how a ROI x traverses it. The traversal procedure is recursive and describes whether a ROI in a node n should go to the left child n_L or to the right one n_R. In detail, in [9] two traversal modalities are defined:

- Given a node n we have one prototype P, i.e. a ROI, and a threshold on the connectivity strength $\theta \in [\min_{x \in n} connectivity(x, P), \max_{x \in n} connectivity(x, P)]$, if $connectivity(x, P) > \theta$ then $x \longrightarrow n_L$, otherwise $x \longrightarrow n_R$. In other words, x ends up in the left child, along with the prototype, only if their connection is strong enough.
- In a node n two different ROIs, P_L and P_R are chosen as prototypes, which respectively represent the putative left and right child. If $connectivity(x, P_L) > connectivity(x, P_R)$ then $x \longrightarrow n_L$, otherwise $x \longrightarrow n_R$. In other words, the ROI x ends up in n_L if its connection to P_L is stronger than its connection to P_R.

In our context, the first traversal modality seems more suitable. In detail, when using only one prototype, by analyzing how strongly connected is the ROI x to the ROI P, we obtain a partial contribution to the total disconnection degree of x, which computation is the final aim of the proposed methodology.

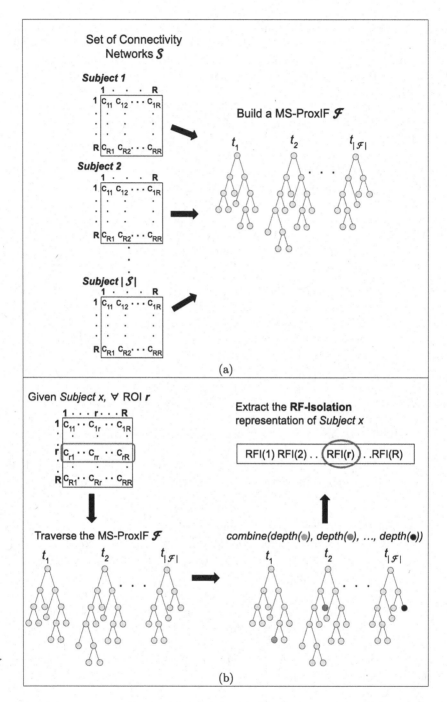

Fig. 1. Pipeline of the proposed approach: (a) building an MS-ProxIF; (b) extracting the *RF-Isolation* vector for a subject x. *Combine* stands for the function computing the anomaly score at forest level via aggregation of the tree scores.

Instead, from the second traversal modality, we can only infer to which prototype the ROI x has a stronger connection, which is not what we are seeking.

The aim behind the building procedure of a ProxIT, is to obtain a tree structure where splitting n generates two child nodes such that: each child contains objects highly similar to each other but dissimilar to the objects in the sibling node. In our context, the aim is to find one child node that contains many ROIs strongly connected with each other, and the other child that contains few ROIs. The former will be the root of a big subtree, and thus the contained ROIs will be isolated after a long procedure. The converse holds for the ROIs in the smaller sibling: they are likely to be isolated soon, i.e. of being disconnected. Ideally we would like a split of a node n to generate a node containing only inliers, highly connected ROIs, and one containing only outliers, disconnected ROIs.

To build a tree, we have to define a learning strategy, i.e. how to define on a node n the test which induces its splitting into two child nodes. In an MS-ProxIT, it consists of choosing: one prototype ROI P, a threshold θ on the connectivity strength with respect to P (or two prototypes P_L and P_R[2]), and in some cases, additionally with respect to [9], a connectivity network. Indeed, an MS-ProxIT can be built with one or several connectivity networks. In the first case, the learning algorithm is identical to [9]: given n, P and θ we split the ROIs in n, based on how strongly connected they are to P. In the latter scenario, instead, since the MS-ProxIT is built using multiple connectivity networks, the test on node n is characterized by P, θ and one connectivity network cn. In other words, to split the ROIs in n, we evaluate how strongly connected they are to P in the chosen connectivity network cn. It is important to note that: i) only ROIs traverse the nodes; ii) the connectivity network cn is needed *only during* the tree building procedure. Therefore, when the latter has ended, we can discard the information about which cn was used to partition the ROIs in n.

The other crucial step of the building procedure consists of choosing the best test for each node n: the choice can be either completely random or based on an optimization procedure. This step is independent of the number of connectivity networks used to build the tree.

In detail, to the already extensive pool of training criteria proposed in [9], i.e. how to choose the best test, we add two additional ones, *O-1PRD* and *O-2PRD*. These criteria are an adaptation to the outlier detection context of the *RényiD* [2] criterion proposed for clustering, i.e. the best split is the one maximizing the divergence in terms of information (quantified by the Rényi entropy) between the two child nodes. This concept is relevant also for outlier detection, since finding the two child nodes conveying the highest amount of different information, corresponds to ideally separating inliers from outliers. In detail, to define these criteria, we estimate the Rényi divergence between the child nodes using an estimator [10] which employs only information related to the K-Nearest Neighbors (KNNs) of the objects. Formally, given a set A of N

---

[2] For the sake of clarity the remainder of the explanation will refer to tests characterized by P and θ, but an analogous reasoning would hold if the test consisted of choosing two prototypes P_L and P_R.

objects and a set B of M objects, the estimation of the Rényi divergence of order α of A from B is computed as follows:

$$RD(A, B) = \frac{1}{\alpha - 1} \log \left[\frac{\frac{M}{N}^\alpha}{M} \sum_{i=1}^{M} (\frac{N_i}{M_i + 1})^\alpha \right] \tag{1}$$

where N_i and M_i are the number of KNNs of $b_i \in B$ that respectively belong to A and B. In other words, A will diverge more from B, if few of its objects are KNNs of objects in B. In our context, A and B are the child nodes n_L and n_R, and to measure their divergence, we have to compute both $RD(n_L, n_R)$ and $RD(n_R, n_L)$ –and average their results– since the divergence is not symmetric. Please note that *O-1PRD* and *O-2PRD* differ only because the former evaluates tests defined by P and θ, whereas *O-2PRD* evaluates tests defined by two prototypes P_L and P_R. For additional information related to the learning phase, see [9].

Step 2. Given a built MS-ProxIT, we can extract, for a given subject and a given ROI x, its anomaly score. Due to their outlier-like nature, disconnected ROIs are more likely to be isolated sooner in the tree, and therefore we would like to assign them a higher anomaly score. To assign such score, we make the ROI x traverse the tree, following one of the two modalities previously presented, depending on the training criterion used to build t (one or two prototypes). It is important to highlight that traversing t is independent of which and how many connectivity networks were used to build the MS-ProxIT. In other words, the evaluation of whether a ROI x should follow the left or right edge depends exclusively on the connectivity network of the subject that is traversing t.

After traversing t, we can recover the anomaly score of the ROI x, which is a function of the depth of the reached leaf. In detail, the smaller the depth, the higher the anomaly score, i.e. the more likely the ROI x is highly disconnected. To obtain the anomaly score of x at forest level, we have to make x traverse each MS-ProxIT composing the MS-ProxIF: the final score is a function of the average depth of the reached leaves –see [8] for details on the formula.

By repeating this procedure for all ROIs of a subject, we obtain the *RF-Isolation* vector, a novel representation where each feature represents the degree of disconnection of a ROI as computed from MS-ProxIF.

Please note that an *RF-Isolation* vector, extracted from an MS-ProxIF – which we recall is trained on the MS population– is an adequate representation also for the healthy cohort. Indeed, given a ROI disconnected only in MS subjects, in a healthy patient the same ROI is likely to traverse a longer path, i.e. get a lower anomaly score, since its connectivity to other ROIs is probably higher.

2.4   Classification Analysis

To evaluate the suitability and robustness of the proposed representation, we made two different analyses: the first studies the MS-ProxIF model, whereas the second one compares *RF-Isolation* to standard network-derived metrics.

Both analyses are based on the classification of the 79 subjects as either MS or healthy, following a Leave One Out protocol (LOO): each classifier is trained 79 times on a dataset of 78 subjects and tested on the left out subject. The classifiers we employed are: linear Support Vector Machines (SVM), K-Nearest Neighbor (KNN) and Random Forest (RF). We measured the classification performances using the Matthews Correlation Coefficient [3] (MCC) which takes into account all four entries of the confusion matrix and thus returns a high value only if both classes are well identified. In detail, we employ the normalized MCC, which ranges in $[0, 1]$.

To increase the robustness of the proposed methodology, we repeat 5 times the whole procedure, i.e. from building the MS-ProxIF to the classification step.

Analysis of the MS-ProxIF Model. The first analysis aims at finding the most suitable variant of the MS-ProxIF model for the task of extracting RF-Isolation–and subsequently of discriminating MS subjects from the healthy cohort. In detail, we generated many different MS-ProxIF models, from which we extracted different *RF-Isolation* vectors, by varying the following parameters:

- Number of trees in a forest: $T \in \{50, 100, 200\}$.
- Number of ROIs used to build each tree: $S \in \{50\%, 75\%, 100\%\}$.
- Number of connectivity networks used to build each tree: $C \in \{1, 5, 10, 20, All\}$ where *All* consists of using all MS subjects to build each tree.
- The training criterion. In detail, we study 6 criteria: *R-1P* and *R-2P* which split data randomly, chosen because of the pervasive success of random variants in [8]; $O\text{-}1PS_D$ and $O\text{-}2PS_D$, the best variant according to [9]; and *O-1PRD* and *O-2PRD*, the novel variants proposed in Sect. 2.3.
- The maximum depth that each tree can reach: $D \in \{\log_2(S), S - 1\}$.

We analyze the behaviour of each of the above parameters via a statistical analysis of the classification results. The input of each analysis is the set of MCC values averaged across the 5 iterations of the proposed approach and all parameters except: the 6 training criteria, the 3 classifiers (SVM, KNN, RF) and the parameter under analysis. For example, if we were to analyze T, the input would be made of 54 MCC values since: we average across the iterations and all values of D, S and C; we have 3 values of T, 3 classifiers and 6 training criteria.

As to the adopted statistical procedure, when analyzing D, since it can assume only two values, we perform a Wilcoxon signed-rank test, whereas for all other parameters, we carry out a Friedman test followed by a post-hoc Nemenyi test. Indeed, the Friedman test is adopted if we have to assess whether there is a global significant difference among more than two methods, in our scenario represented by the different values a parameter can take. Following the Friedman test, we perform a Nemenyi test that employs a critical value to find out which pairs of methods are statistically different. We use a critical difference (CD) diagram [5] to visualize the results of these tests. A CD diagram consists of a line where methods are represented from left to right based on their rank, from worst to best. If two methods are not significantly different, a red line connects them. The significance level is set to $\alpha = 0.05$ for all tests.

Comparison to Standard Network Measures. The second analysis aims at understanding whether *RF-Isolation* has a higher capability of correctly identifying MS subjects compared to standard network-derived measures. In detail, from each subject's connectivity network, we extracted the following metrics using the Brain Connectivity Toolbox [13][3]:

- *Node Strength*: It is the sum of the weights of all edges connected to the node.
- *Local Efficiency*: It is the average inverse shortest path length in the neighborhood of the node; a stronger connection leads to a stronger contribution.
- *Assortativity*: It is a correlation coefficient based on the Node Strength. A positive value indicates that nodes tend to link together with nodes which Node Strength is similar.
- *Clustering Coefficient*: It is the sum of the clustering coefficients of all nodes. The Clustering Coefficient of a node consists of the proportion of its neighbors which are neighbors of each other.
- *Density*: It is the ratio between the number of existing connections to the number of possible connections.
- *Global Efficiency*: It is the average inverse shortest path length in the network.
- *Mean Strength*: It is the average Node Strength computed across all nodes.
- *Modularity*: It is a statistic describing to which degree the network can be partitioned into disjoint sets of nodes, such that within each set the number of edges maximized, and the number of edges between different groups is minimized.

Node Strength and Local Efficiency are local measures, i.e. analogously to *RF-Isolation*, they have one feature for each ROI. Instead, all the remaining metrics are global measures, i.e. one value describes a connectivity network. These measures were chosen because of their significance with respect to MS, as shown in previous works ([15] and references therein). As mentioned in Sect. 1, all the listed metrics are subject-wise, being computed independently for each subject.

3   Results

3.1   Analysis of the MS-ProxIF Model

Table 1 depicts the results of the analysis of the depth D: we report the mean rank of both values of D, and the p-value output by the test, in **bold** if the null hypothesis is rejected. We can assess that using a smaller depth, $D = \log_2(S)$, is the best significant option. The latter is a common choice in the context of isolation-based approaches [8], and it also represents the average tree height [7]. The second parameter we analyze is the forest size T, which results are depicted via the CD diagram in Fig. 2(a): even though $T = 100$ is first-ranked, it is comparable to $T = 200$. Nevertheless, $T = 100$ remains a wiser option

---

[3] For more thorough descriptions, please refer to https://sites.google.com/site/bctnet/list-of-measures.

Table 1. Results of the Wilcoxon signed-rank test comparing the two options for the maximum depth D.

Mean rank		
$D = \log_2(S)$	$D = S-1$	p-value
1.222	1.778	**0.0074**

from a computational perspective, and it is also a common choice for RF-based methodologies [8,12]. Then we analyze the number of ROIs S used to build each tree, which is expressed in percentage 50% = 43 ROIs, 75% = 64 ROIs, and 100% = 85 ROIs. From the CD diagram depicted in Fig. 2(b) we can observe that using all ROIs is the best choice, even though it is comparable to $S = 50\%$. One of the most interesting parameters is C, the number of connectivity networks used to build each tree. The results of the statistical analysis, shown in Fig. 2(c), assess that $C = 10$ is the best option, even though it is comparable to all other values of C, except for $C = 1$, the last ranked. Results confirm the usefulness of employing multiple connectivity networks, i.e. $C > 1$, to build each tree, since it leads to a more informative *RF-Isolation* representation. The last analysis compares the different training criteria: the CD diagram in Fig. 2(d) assesses that *O-1PRD* is the best choice. Further, training criteria based on two prototypes are all ranked significantly worse than all criteria using one prototype, making the latter a better choice, as hypothesized.

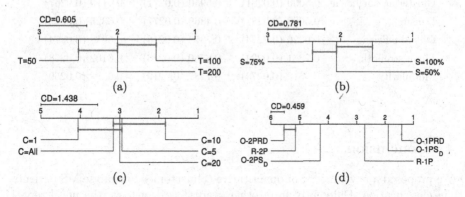

Fig. 2. CD diagram comparing the different options for: (a) the forest size T; (b) the number of ROIs S used to build each tree; (c) the number of connectivity networks C used to build each tree; (d) the adopted training criterion.

Summarizing, we can conclude that even though we can extract a robust *RF-Isolation* vector from several parametrizations, we can identify one that is the most suitable. In detail, we should set the parameters of *MS-ProxIF* as follows: $D = \log_2(S)$, $S = 100\%$, $C = 10$, $T = 100$, and the training scheme to *O-1PRD*.

3.2 Comparison to Standard Network Measures

In Table 2 we report the comparison, in terms of MCC, between standard network-derived measures, and the proposed approach. As to *RF-Isolation*, the MCC has been averaged across the iterations, and the parameters of the underlying MS-ProxIF model have been set according to the previous analysis. We can observe that *RF-Isolation* shows better performances than all the other measures independently of the employed classifier. The only exception is when we compare to Local Efficiency using KNN: however, the difference is rather small with respect to the improvements brought by *RF-Isolation*. To validate the improvement, we also report the standard errors of the mean.

Obtained results suggest that *RF-Isolation*, a disease-wise representation extracted from a trained model, represents a valid alternative to standard metrics for connectivity networks, which are instead subject-wise.

Table 2. Comparison of RF-Isolation with standard network-derived metrics.

Representation	MCC		
	SVM	KNN	RF
RF-Isolation	**0.8686 (0.0128)**	0.7822 (0.0192)	**0.8554 (0.0139)**
Local efficiency	0.7997 (0.0180)	**0.7843 (0.0190)**	0.7799 (0.0193)
Nodal strength	0.8167 (0.0168)	0.6559 (0.0254)	0.8127 (0.0171)
Assortativity	0.5000 (0.0281)	0.5000 (0.0281)	0.4790 (0.0281)
Clustering coefficient	0.5000 (0.0281)	0.5950 (0.0271)	0.6110 (0.0267)
Density	0.5000 (0.0281)	0.4369 (0.0277)	0.6325 (0.0262)
Global efficiency	0.6226 (0.0264)	0.7373 (0.0218)	0.5818 (0.0274)
Mean strength	0.7261 (0.0224)	0.6760 (0.0246)	0.6102 (0.0268)
Modularity	0.5812 (0.0274)	0.7487 (0.0212)	0.6392 (0.0259)

4 Conclusion

We proposed a novel metric of quantitative connectomes to analyse MS patients via classification. Differently from other graph-based metrics, the novel representation, *RF-Isolation*, is more descriptive from a disease point of view. Indeed, to extract it, we trained a model on the entire MS population, from which we derived the *RF-Isolation* vector for all subjects. The suitability of the approach is confirmed by the experimental analyses. Future work includes analyzing the link between *RF-Isolation*, and the most involved ROIs in the MS lesion process. In addition, when a bigger dataset will be available, we will test *RF-Isolation* for classifying different subtypes of MS, which is nowadays clinically challenging. Lastly, we could also adapt the proposed methodology to study other neurodegenerative diseases, e.g. Amyotrophic Lateral Sclerosis.

Acknowledgments. The authors declare that there is no conflict of interest. This work was partly supported by grants from NMSS (RG 5120A3/1) and Teva Neuroscience (CNS-2014-221).

References

1. Bester, M., Petracca, M., Inglese, M.: Neuroimaging of multiple sclerosis, acute disseminated encephalomyelitis, and other demyelinating diseases. In: Seminars in Roentgenology, vol. 49, pp. 76–85 (2013). https://doi.org/10.1053/j.ro.2013.09.002
2. Bicego, M.: Dissimilarity random forest clustering. In: 2020 IEEE ICDM, pp. 936–941 (2020). https://doi.org/10.1109/ICDM50108.2020.00105
3. Chicco, D., Jurman, G.: The advantages of the Matthews correlation coefficient (mcc) over f1 score and accuracy in binary classification evaluation. BMC Genomics **21**(1), 1–13 (2020). https://doi.org/10.1186/s12864-019-6413-7
4. Daducci, A., Dal Palù, A., Lemkaddem, A., Thiran, J.P.: COMMIT: convex optimization modeling for microstructure informed tractography. IEEE Trans. Med. Imaging **34**(1), 246–257 (2014). https://doi.org/10.1109/TMI.2014.2352414
5. Demšar, J.: Statistical comparisons of classifiers over multiple data sets. J. Mach. Learn. Res. **7**, 1–30 (2006)
6. Desikan, R.S., et al.: An automated labeling system for subdividing the human cerebral cortex on MRI scans into gyral based regions of interest. Neuroimage **31**(3), 968–980 (2006). https://doi.org/10.1016/j.neuroimage.2006.01.021
7. Knuth, D.E.: The art of computer programming. Sorting Search. 3, Ch–6 (1973)
8. Liu, F.T., Ting, K.M., Zhou, Z.H.: Isolation-based anomaly detection. ACM Trans. Knowl. Discov. Data **6**(1), 1–39 (2012). https://doi.org/10.1145/2133360.2133363
9. Mensi, A., Bicego, M., Tax, D.M.: Proximity isolation forests. In: 2020 25th ICPR, pp. 8021–8028. IEEE (2021). https://doi.org/10.1109/ICPR48806.2021.9412322
10. Noshad, M., Moon, K.R., Sekeh, S.Y., Hero, A.O.: Direct estimation of information divergence using nearest neighbor ratios. In: 2017 IEEE International Symposium on Information Theory (ISIT), pp. 903–907. IEEE (2017)
11. Pagani, E., et al.: Structural connectivity in multiple sclerosis and modeling of disconnection. Mult. Scler. J. **26**(2), 220–232 (2020). https://doi.org/10.1177/1352458518820759
12. Probst, P., Boulesteix, A.L.: To tune or not to tune the number of trees in random forest. J. Mach. Learn. Res. **18**(1), 6673–6690 (2017)
13. Rubinov, M., Sporns, O.: Complex network measures of brain connectivity: uses and interpretations. Neuroimage **52**(3), 1059–1069 (2010). https://doi.org/10.1016/j.neuroimage.2009.10.003
14. Schiavi, S., et al.: Classification of multiple sclerosis patients based on structural disconnection: a robust feature selection approach. J. Neuroimaging (2022). https://doi.org/10.1111/jon.12991
15. Schiavi, S., et al.: Sensory-motor network topology in multiple sclerosis: structural connectivity analysis accounting for intrinsic density discrepancy. Hum. Brain Mapp. **41**(11), 2951–2963 (2020). https://doi.org/10.1002/hbm.24989

Summarizing Global SARS-CoV-2 Geographical Spread by Phylogenetic Multitype Branching Models

Hao Chi Kiang[1(✉)], Krzysztof Bartoszek[1], Sebastian Sakowski[2], Stefano Maria Iacus[3], and Michele Vespe[4]

[1] Department of Computer and Information Science, Linköping University, Linköping, Sweden
{hao.chi.kiang,krzysztof.bartoszek}@liu.se
[2] Faculty of Mathematics and Computer Science, University of Łódź, Łódź, Poland
sebastian.sakowski@wmii.uni.lodz.pl
[3] Institute of Quantitative Social Sciences, Harvard University, Cambridge, MA, USA
siacus@fas.harvard.edu
[4] European Commission, Joint Research Centre, Ispra, Italy
michele.vespe@ec.europa.eu

Abstract. Using available phylogeographical data of 3585 SARS–CoV–2 genomes we attempt at providing a global picture of the virus's dynamics in terms of directly interpretable parameters. To this end we fit a hidden state multistate speciation and extinction model to a pre-estimated phylogenetic tree with information on the place of sampling of each strain. We find that even with such coarse–grained data the dominating transition rates exhibit weak similarities with the most popular, continent–level aggregated, airline passenger flight routes.

Keywords: COVID-19 · Hidden Markov model · Phylogeography · State-dependent diversification

1 Introduction

Following an initial outbreak in Wuhan, China at the end of 2019, the severe acute respiratory syndrome coronavirus 2 (SARS–CoV–2) started spreading rapidly around the world in early 2020. It is reported in the literature that asymptomatic and mild cases can make up over 95% of all SARS–CoV–2 infections [22], therefore, one can easily imagine silent transmissions across countries and continents, even with multiple stepping stones, being undetected by health service monitoring. Such transmissions, nonetheless, will leave information in the viral genomes and hence also could be present in the phylogeny connecting the strains.

Supplementary Information The online version contains supplementary material available at https://doi.org/10.1007/978-3-031-20837-9_14.

D. Chicco et al. (Eds.): CIBB 2021, LNBI 13483, pp. 170–184, 2022.
https://doi.org/10.1007/978-3-031-20837-9_14

Some phylogenetic or stochastic-branching-process-based studies which link SARS–CoV–2 genomes to geographical origins have been done previously. For instance, [23] proposed two branching process models which use only the daily laboratory-confirmed case count. There are also phylogenetic studies that focus on tracing the transmission pattern on a local scale. To name a few, [6] has analyzed the local transmission pattern in Aotearoa, New Zealand by comparing a local phylogeny, which consists of 649 virus genomes sampled between February and May, 2020, with the global COVID–19 genome data; [10] did a phylogenetic ancestral trait reconstruction analysis for Boston, Massachusetts, United States; [21] uses the NextStrain data to perform a phylogenetic inference for Ontario, Canada; and [14] for Austria.

Global transmission, however, can be statistically more challenging to infer using phylogenies constructed from virus genomes as the sole material, partly because of the fact that even a genomic sample consisting of tens of thousands data points only represents a tiny fraction of the underlying population. Nonetheless, in an analysis done on a sample of global genomes, [12] observed that "transmission occurred more readily among neighboring countries than among countries that are geographically disconnected", and sequences from China, Hong Kong, and Taiwan tend to form a cluster; so do those from Sweden, Denmark, and Finland. Their observation indicates that it is indeed possible to detect signals of transmission even if the sample size is small on a global scale. Unlike between-country transmission, however, intercontinental transmission should be less correlated to geographical proximity, as one may reasonably suspect that the virus transmits more frequently between North America and Europe than between North America and Africa due to their econo-political kinship, while it is hard to say which

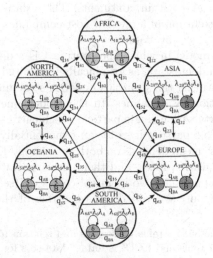

Fig. 1. Graphical representation of the HiSSE model and its speciation and transition parameters. Collapsing the hidden states will result in the multistate speciation and extinction model (MuSSE, [4]) model. More discussion on what the hidden state captures can be found in Sect. 3. We call λ the "diversification rates" and q the "transition rates".

pair of the continents are geographically closer. We ponder whether this kind of correspondence is observable in a global, intercontinental, genome data set.

To investigate whether such signal is indeed detectable on an intercontinental-scale phylogeny, in this work, we attempted to fit the hidden-state multi-state speciation and extinction (HiSSE, [2]) models to a continent-labelled phylogeny that is constructed using global virus genomes, and compared it with intercontinental air passenger volume. We found that the "big-picture" transmission pattern recovered from the phylogeny seems to exhibit a connection to air traffic volume.

The readers are reminded that the question under our consideration is not whether the intercontinental flights are indeed correlated with the disease transmission, but rather, whether a noisy continent-level phylogeny contains signal of such transmission or not. The answer to the former question is, of course, widely expected to be affirmative, while the latter is much less investigated as most studies of this type uses only a local-level phylogeny.

2   Data and Methods

We obtained a phylogenetic tree of the SARS–CoV–2 virus with 3585 tips, dated on 26th April 2020 from NextStrain [7]. Each tip is labelled with one of the six continents: Africa, Asia, Europe, North America, Oceania, and South America. Table 1 shows the number of strains for each continent and the corresponding fatality counts. Among multiple helper functions and tools, the R [16] packages, "ape" [13] and "phytools" [17] were used to pre-process and experiment with the phylogenetic tree.

We make use of the SABRE data set [18] to estimate the bi-directional flows of passengers between and within continents. These data cover almost all air traffic data with the exclusion of a few low-cost companies, and is by far one of the most complete data sets available.

We used a HiSSE model in which the continent is a discrete observed trait and there is a binary hidden-state, as graphically illustrated in Fig. 1. The model was developed for capturing the scenarios in which some unobserved factors are influencing the evolutionary dynamics. In the model, one can think that the branching process starts with a single particle which carries an hidden state and a non-hidden state. This particle lives for an exponentially-distributed random duration of time, whose mean depends on both the non-hidden and hidden state's diversification and transition rates, and then, either mutates into another type with different states, splits into two identical copies of itself that behave in the same manner, or dies out. We will discuss the role of the hidden state later

Table 1. Division of the 3585 strains by regions and relation to observed COVID–19 deaths as of NextStrain tools and ECDC's data—data sets for the day of 26th April 2020 (204494 case fatalities).

	Africa	Asia	Europe	N. America	Oceania	S. America
No. strains (%)	88 (2.45)	610 (17.02)	1641 (45.77)	971 (25.09)	186 (5.19)	89 (2.48)
Thous. deaths (%)	1.38 (0.67)	16.87 (8.25)	123.30 (60.30)	57.12 (27.93)	0.11 (0.05)	5.72 (2.80)

Table 2. Parameter constraints used in the analyses. All models assume that the two hidden–state transition rates are equal and that the extinction rates of each region are equal. The indices i, j correspond to the six possible regions. BIC stands for the Bayesian information criterion and AIC stands for the Akaike information criterion. "All equal" means that the rates of the six continents are assumed to be the same, while "free" means that each continent has its own rate.

	λ_{obs}	q_{obs}	BIC	AIC	Evidence	Entropy
Model I	Free	Free	–	–	–	–
Model II	Free	$q_{ij} = q_{ji}$	**2532.282**	2383.854	**-1206.655**	1180.386
Model III	All equal	$q_{ij} = q_{ji}$	2560.932	2443.426	-1234.223	1213.961
Model IV	All equal	Free	2574.625	**2364.352**	-1230.033	**1164.299**
Model V	Free	All equal	2828.881	2767.036	-1988.705	1384.707
Model VI	All equal	All equal	2864.063	2833.140	-1433.883	1417.864

in Sect. 3; the non-hidden state has six possible values, each corresponds to a continent. The observed and hidden states carry their own diversification and transition rates; and we used the Markov Chain Monte Carlo (MCMC) method implemented in RevBayes 1.0.12 [8] to estimate six similar models, each of which has different parameter space restrictions, as exhibited in Table 2. In the table, "all equal" means that the rates of the six continents are assumed to be the same, while "free" means that each continent has its own rate. Extinction rates are present in the model but not in the graph for readability. We tried different restrictions in order to improve estimability and reduce the risk of parameter redundancy, as determining parameter identifiability is not a straightforward task even for the simpler non-phylogenetic hidden Markov models that do not have direct transitions between observed states [3].

Note that we call q's the "transition" rates, while the word "transmission" is reserved to colloquially mean the virus' spreading in a more general sense. It would be rather misleading to call the q's "transmission rates" as they only capture the intercontinental dynamics, while "transmission rate" often suggests the speed of the virus' spreading from one individual to another regardless of whether this spreading is intercontinental or not, as in, for example, the popular susceptible-infectious-recovered (SIR, [9]) epidemic models.

For each model, 32 independent chains are computed in parallel and stopped simultaneously. Then we trim them to the same length (number of cycles, as in RevBayes each such iteration actually contains multiple proposals) and discard a quarter of our sample as burn-in. The resulting chain lengths of each model are $(257, 1103, 928, 946, 1103, 1103)$. Model I, the unconstrained model, has a problem that the 32 chains disagree with each other, with some of the chains running very slowly, hence the short chain length. We discontinued running this model due to the huge running time and limited computation budget, as a single run of this set of experiments could take at least about 5000–7000 processor hours to complete.

The prior of the root state, in our case, is set to a discrete distribution with 47.5% probability on each of Asia–A and Asia–B, while the remaining 5% probability mass is equally distributed over all the other states; this reflects our knowledge that the first outbreak happened in China, and that we do not know the hidden state at the root. Due to the high dimensionality of the estimation problem, we impose a constraint that the transition rates between the two hidden states are equal, that is, in Fig. 1, we have $q_{AB} = q_{BA}$. Except for the mentioned and a few minor constants, we follow the recommendations given by RevBayes' tutorial (https://revbayes.github.io/tutorials/sse/hisse.html) on the HiSSE model and did not add any meaningful information to the prior.

After obtaining the posterior distribution of the diversification and transition rates, we compared their posterior means with the percentage of intra- and inter-continental travel attributed to each continent.

To compare the models, we have computed the Bayesian Information Criteria (BIC, [20]) , Akaike information criterion (AIC, [1]) , estimated evidence and an estimate of the posteriors' differential entropy. The BIC and AIC are computed using the posterior mode in the approximately 32 thousand MCMC sample points. The evidence is estimated using the harmonic mean estimator [11]. The differential entropy is estimated by taking the mean of the log-posterior density across the MCMC sample points.

We have also attempted to constrain the extinction rate to zero before running the six main models, but with much shorter MCMC chains. The resulting parameter estimates, although only from around 200–300 MCMC sample points, are all very similar to the main model, and this coincides with the fact that the main models' extinction rate estimates are all very small (on the scale of 0.001). In other words, the two sets of models appear to agree on the extinction rate, although it has been observed, e.g. in [2], that extinction rates are rather difficult to estimate in general. We did not run these models further as there seems not to be much insight we can get from them, and running the MCMC chain is computationally expensive.

In addition to the six ways of restricting the parameters as shown in Table 2, which are our main models, we have also experimented with a version of each of them, except Model VI (because its speciation and transition rates are all equal, so this sort of prior would not bring in anything), in which the SABRE data were used as a weakly informative prior rather than as an independent piece of information to compare the fit parameters with. The prior-informed models are essentially the same as their original counterparts, except that (1) on the fractions $[\lambda_1, \cdots, \lambda_6] / \sum_{j=1}^{6} \lambda_j$, whenever its six dimensions are not restricted to be the same, a six-dimensional Dirichlet prior with $\alpha = c_1 [\ell_1, \cdots, \ell_6]$ is imposed, where ℓ_i is the percentage of intra-continental flight traffic attributed to the ith continent; and likewise, (2) on the fractions $[q_1, \cdots, q_p] / \sum_{j=1}^{p} q_j$, in which p is either 15 or 30 depending on whether or not $q_{ij} = q_{ji}$ is assumed in Table 2, whenever its p dimensions are not restricted to be the same, a p-dimensional Dirichlet prior with $\alpha = c_2 [\kappa_1, \cdots, \kappa_p]$ is imposed, where the κ_i's are the percentages of inter-continental flight traffic attributed to all possible pairs of continent. The λ's informative Dirichlet prior is applied to Model I, II, and V because those are the models whose λ's are not restricted to be all equal. Similarly, the mentioned informative prior for q is applied to only Model I–IV as Model V and VI have their q's equality-constrained. This prior is inapplicable to Model VI because neither its λ's nor q's are allowed to vary across continents. In other words, we expect the percentages

$\lambda_i / \sum_{j=1}^{6} \lambda_j$ and $q_i / \sum_{j=1}^{p} q_j$, for each continent i, to have their expected values equaling to ℓ_i and κ_i, respectively, and both ℓ_i and κ_i are taken from the SABRE data. We have chosen $c_1 = 6.0$ and $c_2 = 1.0$ in our experiment. The reason why c_2 is set so much lower than c_1 is that some κ_i in the SABRE data are much closer to zero than others. The marginal prior distributions of the dimensions whose κ_i's are closer to zero will have a much lower variance than those whose κ_i are further from zero. Therefore, had we not relaxed c_2, the Dirichlet prior would have been probably too restrictive on some dimensions.

3 Results and Discussion

The MCMC chains of Model I has failed to converge in a sense that the 32 chains disagree with each other and the sampling of some chains is very slow, while the chains of all other models have Gelman-Rubin's $\hat{R}$ statistics [5] well below 1.1 (See Figure S.1 in the Supplementary Information). Model I's problem persists with varying random seeds and random initializations. Given that both Model IV (with free transition rates) and Model V (with free diversification rates) converge nicely, we suspect that it is the interaction between the λs and the qs that makes the geometry of the likelihood function difficult to sample from.

Tables 3, 4, and 5 show a comparison between our Model II and IV's rate estimates and the air traffic data. Figure 2 depicts our marginal posterior. When reading the mentioned tables, the readers should note that our main focus when comparing the rates with the passenger flight volumes is not on the actually estimated values, but the proportion of transitions between a particular pair of continents to the total transition volume.

According to the Bayesian Information Criterion (BIC) and the estimated evidence, the best model was Model II, in which all diversification rates, observed and hidden, are allowed to vary freely but the transition rates "there–and–back" between each pairs of regions are assumed equal, in other words, we have $q_{ij} = q_{ji}$. We found reasonable that a symmetric-rate model was chosen, as the intercontinental air traffic pattern is rather symmetric according to our data set. However, the AIC and the entropy favours Model IV, which constrains all diversification rates to equality but leaves the transition rates free. It is noticeable that Model II gives us narrow credible intervals for the transition rates, but wide intervals for the diversification rates; while Model IV behaves contrariwise, with some of its estimated transition rates having large uncertainty but diversification rates well-estimated. In fact, the average of all six diversification rates of Model II almost coincides with Model IV's estimate. The diversification rates are orders of magnitude larger than nearly all of the estimated intercontinental transition rates. The transition rates estimated by the two also shows similarities. For example, from Fig. 2 it seems both models agree that the Europe-North-America rate is around 0.3, Asia-North-America slightly less than 0.1, and Asia-Africa's rate is small.

Model II and IV almost completely agree on the rates related to the hidden states. It is worth noting that both estimated λ_A to be rather small, meaning

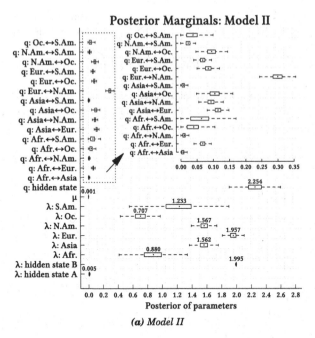

(a) Model II

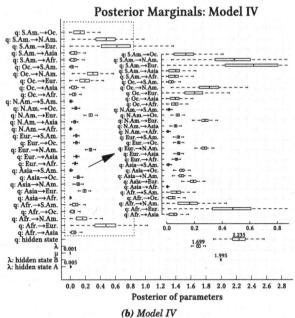

(b) Model IV

Fig. 2. Marginal posterior distributions under Model II and IV , in which λ, q are μ are the diversification rates, transition rates and extinction rates. The arrows in the graphs indicate zoom-in of the content in the dotted boxes. Numbers above boxes indicates posterior mean.

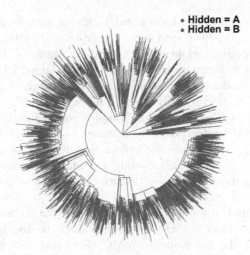

Fig. 3. Stochastic character map for the hidden state. Note that state A's branches are significantly longer.

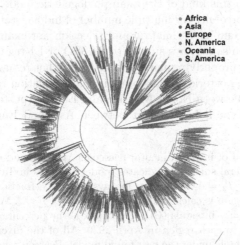

Fig. 4. Phylogeny of SARS–CoV–2 strains from nextstrain.org [7,19] with maximum a posteriori geographical states of the lineages painted. The unit of time of the tree is 30 days. The maximum height of the tree is 137 days. The ancestral regions for the strains are found under the best found Model II.

that the internal nodes in latent state A will have much slower speciation. That is to say, the latent state A has explained the very long branches in the phylogeny (See Fig. 3). In fact, we have previously attempted to fit, using the R package "diversitree" [4], a multistate speciation and extinction model (MuSSE), which is similar to the HiSSE but without the hidden states, as it seems to us that a natural model to use should be one that takes the speciation, extinction and transition rates to be region-dependent. Unfortunately, without the hidden states,

multiple optimization routines have failed to find a maximum likelihood estimate, in a sense that the likelihood seems to tend toward infinity, as some rates
move toward infinistesimal while others toward big numbers. That λ_A captured
the excessively long branches may have, in part, explained why the estimation
is stablized after introducing the latent states, because the long branches may
be explainable under MuSSE only if some λ's are tiny. The results given by
MuSSE can be found in Table S.1 in the Supplementary Information. We do not
know exactly what the latent state is, in other words, what has caused some
branches to be long; nonetheless, we believe that having latent states is really
not as peculiar a choice as it initially might seem, for claiming that no latent
factors are affecting the transmission pattern can be just as questionable, if not
more, in such a global pandemic.

On the other hand, a more complicated model (for instance with multiple
hidden states rather than a binary) would perhaps be too heavy or hard to
estimate given that the full model's 32 MCMC chains have problem agreeing
with each other. The choice of the optimal number of hidden states is, in fact,
an interesting and rather unexplored model selection problem, as given the high
computation cost in this kind of Bayesian phylogenetic models, it would not be
practical to brute-forcefully try multiple number of hidden states. Thus, one may
need a method to combine estimation and selection, for example by employing
some clever priors in some ways similar to LASSO or L1-regularized regression.
But how to do this properly remains a topic for further research.

In Model II, the posterior means of almost all transition rates are below 0.1
except for the one between North America and Europe (posterior mean=0.305,
the highest) and, next below it, that between Asia and Europe (0.115). The

Table 3. Comparison of intracontinental passenger volumes and diversification rates
λs. The Roman numeral subscript indicates model numbers in Table 2. The posterior
mean doubling time, $T_2 = 1/\lambda$, is given in days. The %λ, fraction of diversification,
attributed to each region is calculated for region i as $\lambda_i / \sum_j \lambda_j$. We also present their
95% posterior credibility intervals (CI). The %pax is the percentage of intracontinen-
tal travel attributed to each region in April 2020. All of the diversification rates are
from the posterior sample under the best found model. Pearson's correlation coefficient
between $E\%\lambda$ and %pax (see Fig. S.20 in the SI for a visualization) is 0.425 (p–value
0.4) and Kendall's τ is 0.733 (p–value 0.056).

	Africa	Asia	Eur.	N.Am.	Oc.	S.Am.
$E\lambda_{II}$	0.880	1.562	1.957	1.567	0.707	1.233
$E\lambda_{IV}$			1.699			
$ET_{2,II}$	35.005	19.21	15.293	19.334	42.92	25.717
$ET_{2,IV}$			0.589			
$E\%\lambda_{II}$	11.1	19.8	24.8	19.9	8.9	15.6
$\%\lambda_{II}$ 95%CI	8.0	18.0	22.9	18.1	7.0	11.3
	14.6	21.6	26.7	21.6	11.0	20.2
%pax	1.9	69.2	13.0	11.8	1.5	2.6
%deaths	0.67	8.25	60.30	27.93	0.05	2.80

Table 4. Comparison of passenger volume proportions and transition rates estimated from Model II. The first two rows of each cell correspond to the estimated fraction of the transitions that is attributed to the given pair of regions, e.g. for regions 1 and 2 it will be $q_{12}/\sum_{i<j} q_{ij}$. In each cell, we have in the top row the posterior 95% credibility intervals, then the posterior average and in the third row the fraction of April 2020 intercontinental air passenger travel attributed to the given pair of regions. Pearson's correlation coefficient between the passenger fractions and transition rate fractions is 0.67072 (p-value 0.0062) and the Kendall's τ is 0.35238 (p-value 0.074).

	Asia	Europe	N. America	Oceania	S. America
Afr.	(0,0.015)	(0.046,0.076)	(0.003,0.019)	(0.006,0.080)	(0.008,0.127)
	0.006	0.060	0.009	0.035	0.056
	0.052	0.139	0.012	0.001	0.002
Asia		(0.083,0.123)	(0.063,0.106)	(0.066,0.125)	(0.002,0.018)
		0.103	0.083	0.094	0.008
		0.297	0.086	0.048	0.004
Eur.			(0.232,0.302)	(0.058,0.097)	(0.046,0.078)
			0.267	0.076	0.061
			0.244	0.014	0.044
N.Am.				(0.061,0.113)	(0.010,0.035)
				0.086	0.021
				0.008	0.048
Oc.					(0.007,0.077)
					0.034
					0.001

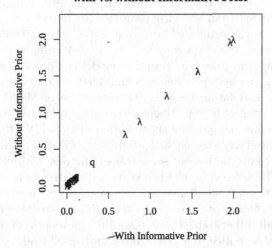

**Model II's Posterior Means
with vs. without Informative Prior**

Fig. 5. The posterior mean estimates of Model II using the SABRE data as an informative prior versus without using it. Each estimated diversification rate is displayed as a "λ" symbol in the plot. Likewise, the transition rate estimates are shown as "q".

Table 5. Comparison of passenger volume proportions and transition rates estimated from Model IV. The format and meaning of each row is the same as Pearson's correlation coefficient between the passenger fractions and transition rate fractions is 0.24277 (p-value 0.196) and the Kendall's τ is 0.24138 (p-value 0.063).

From	To					
	Afr.	Asia	Europe	N. America	Oceania	S. America
Afr.		(0.001,0.033)	(0.038,0.183)	(0.006,0.083)	(0.000,0.031)	(0.002,0.042)
		0.012	0.108	0.037	0.010	0.017
		0.026	0.071	0.007	0.001	0.001
Asia	(0.000,0.004)		(0.031,0.060)	(0.016,0.036)	(0.017,0.035)	(0.001,0.007)
	0.001		0.044	0.025	0.026	0.003
	0.027		0.156	0.045	0.023	0.002
Eur.	(0.012,0.022)	(0.015,0.027)		(0.049,0.079)	(0.014,0.027)	(0.015,0.026)
	0.016	0.020		0.063	0.020	0.020
	0.068	0.141		0.121	0.006	0.021
N.Am.	(0.000,0.006)	(0.009,0.022)	(0.049,0.082)		(0.012,0.027)	(0.001,0.008)
	0.003	0.015	0.064		0.019	0.004
	0.006	0.041	0.123		0.003	0.024
Oc.	(0.004,0.031)	(0.001,0.037)	(0.016,0.086)	(0.035,0.113)		(0.001,0.021)
	0.015	0.015	0.047	0.071		0.009
	0.001	0.025	0.008	0.004		0.000
S.Am.	(0.001,0.042)	(0.001,0.047)	(0.054,0.238)	(0.051,0.185)	(0.003,0.082)	
	0.014	0.016	0.142	0.112	0.032	
	0.001	0.002	0.023	0.023	0.000	

mentioned top two agree with the volume of passenger in the SABRE data set, although the third most busily travelled Africa-Europe pair is not that visible in the transition rates, which, readers are reminded, are estimated purely using the phylogeny constructed from the virus genome. Nonetheless, we have observed correlation between the estimated transition rates and air traffic data. If one performs a Bayesian model averaging on Models II–V using our estimate of the evidence and a uniform prior, the resulting model is almost identical to Model II, whose weight given by the evidence is extremely close to one.

The Pearson's correlation between the proportion of Model II's estimated q's and the proportion of intercontinental passenger volume is 0.67072 (p-value 0.0062), which is much more significant than that of Model IV's (0.24277, p-value 0.196). This is indeed expected as, first, from Fig. 2b we can see that Model IV's posterior is rather uncertain about some transition rates; and secondly, the passenger volume in Table 5 is very close to a symmetric matrix. In fact, the symmetry is very consistent throughout all months of the year, as presented in Figure S.21 in the SI, where one should notice, *en passant*, that the intercontinental traffic proportions are overall rather stable throughout different months of the year. On the intracontinental side, notice that Asia's intracontinental traffic volume is 69.2% of the total (the penultimate row in Table 3), which can severely affect the computed Pearson's correlation (0.425, p-value 0.4, Table 3); but, in fact, if we exclude Asia, then the Pearson's correlation would become 0.92996 (p-value 0.022), and the Kendall's τ would become 1.0 (p-value 0.0167). Of course, however small these Asia-excluded p-values are, they were calculated only from five data points; still and all, the pattern of correlatedness is more apparent when viewed together with the transition rates' correlations (for a visualization of these data points see Figure S.20 in the SI). A closer look into Table 4 reveals that the estimated transition rate proportions and the air traffic volume are much more similar among pairs involving Europe, North America and Asia, and the very dissimilar rates in the table tend to be the pairs between Oceania, South America and Africa. This dissimilarity may very well be the result of the fact that we have significantly less data from the latter group of continents (Table 1).

As mentioned in the method section, we have also attempted using the SABRE data as a prior instead. With the prior's presence, the MCMC sampler struggles to output parallel chains with as good a mixing as the original models, and this has led to that some $\hat{R}$'s in all but Model V being larger than 1.1 (See Figure S.1 in the Supplementary Information). We suspect that the source of this problem is the efficiency of the MCMC sampler. The auto-correlation in the sample was visibly higher than the original versions of the models, and we have observed that the $\hat{R}$ statistics generally improve as more sample points were collected, although the improvement was rather slow, especially in the models in which all the q's are allowed to vary freely.

Although some chains may not have fully explored the entire posterior distribution, Fig. 5 suggests that at least the posterior mean estimates of Model II are rather robust against the added prior. In fact, the Pearson's correlation in Fig. 5 is 0.999924 and Kendall's τ 0.977. Again, the readers are reminded that, although it may not be the case seeing the multidimensional prior as a whole, some of the prior's marginal distributions with even only $c_2 = 1.0$ is in fact quite restrictive, especially in the cases where the inter-continental traffic of a continent pair is tiny. In Model II, another noticeable example of the phylogenetic data overwhelming the prior is the transition rate between Oceania and South America. The original model and the informative-prior version estimated this rate to be 0.034 and 0.029 respectively, while the prior's expected value for this rate is 0.001; i.e., the final estimates are thirty times bigger than the prior, and the prior has only dragged the posterior mean down a little.

It has to be discussed what type of data the nextstrain.org database is providing, in the context of our analyses. Phylogenetic trees constructed by the NextStrain platform are estimated from DNA sequences using an implementation of the FastTree algorithm [15] in the bioinformatics tool called "augur". Therefore how the DNA sequences were sampled, or sub-sampled from a bigger pool of sequences, may affect the estimated tree's topology, as well as the result of our model. One limitation of our model and the data is that the percentages of strains that we have for each continent (See Table 1) may not match the actual "population size" of each continent. For example, Europe accounts for 60.30% of all death while it is only 45.77% of our strains, and this may bias our estimates. Not only do these proportions matter, the distribution of cities from which the genomes have originated may also influence the topology of the estimated tree, as one could easily imagine the virus genome sampled from Europe's important metropolises being different from small cities. The degree to which this variability can influence the estimated tree is rather unknown and difficult to quantify.

The diversification rate λ is related to the expected doubling time, i.e. as $1/\lambda$. Hence, this allows us to present the diversification rates in terms of units of time. The expected posterior times for a strain to produce a new one are presented in Table 3. The transition rates are strongest between regions most experiencing the virus, but also seem to correspond, on a very high–level, to human air travel paths. Three of the four highest transition rates correspond to the most frequent flight paths. Under this kind of interpretation, one could ask why the transition rates are orders of magnitude smaller than the diversification rates. Firstly, only a small percentage of the populations travels intercontinental; and secondly, travel restrictions and border closures may also play a role. Also, we must point out here that the assumption of constant-through-time migration rates between continents only reflect the average rate during this time period. The volume of travel, from a fine-grained view, is expected to fluctuate due to both randomness and travel restriction during the period. Therefore, the estimated q_{ij} parameters should be understood as correlating to the average transition rates for this time period.

4   Conclusions

We found an unsurprising, small, but statistically significant, correlation between the transmission intensity that is estimated from only the molecule-level information and the intercontinental volume of air passengers (Tables 4 and 5). Both of them suggest that North-America-Europe and Asia-Europe are likely to be the top two pairs of continents in transmission intensity during our considered time period.

Though, it has to be underlined at the end that the proposed method does not capture any particular transmission history, infection pathways, nor does it indicate any preventative strategies, decisions, nor policies. The branching processes' parameters are time–homogeneous, so no temporal patterns can be described. It rather provides a general transmission and diversification "big picture" that is averaged over time.

Nonetheless, we find it rather remarkable that, via our novel application of the HiSSE model in such an analysis, we are able to see, at this coarse-grained level, any numerical evidence that one can indeed observe signals of intercontinental transmission solely from a phylogeny that is constructed from molecule-level information of the virus genomes, despite the very sparse and possibly biased sampling process. It would be an interesting further study to investigate the possibility of applying this kind of models to finer-grained phylogenetic data, for example, on a country-level.

Acknowledgements. We thank Fredrik Ronquist for very valuable comments. K.B.'s research is supported by Vetenskapsrådets Grant 2017–04951 and partially by an ELLIIT Call C grant. H.K.'s research is partially supported by Vetenskapsrådets Grant 2017–04951.

Availability of Data and Materials. The R scripts, RevBayes scripts, MCMC chains, along with the used phylogenetic tree, geographical classification, inside and between regions air passenger volume fractions are available at https://github. com/KHDS-mod/COVID-19-HiSSE and https://urn.kb.se/resolve?urn=urn:nbn:se: liu:diva-185867. An already constructed phylogenetic tree and strain (i.e. leaf) data were downloaded from NextStrain (https://nextstrain.org/ncov/global) on 26th April 2020. This data set contains 3585 genomes sampled between December 2019 and April 2020. A full acknowledgments table of the research groups and authors from the whole world generating the sequence data, from which NextStrain's phylogenetic tree is constructed, is provided in the `nextstrain_ncov_global_authors.tsv` file in `COVID-19-HiSSE` repository. The geographic distribution of COVID–19 case fatalities worldwide (presented in Tab. 1) were downloaded from European Centre for Disease Prevention and Control (https://www.ecdc.europa.eu/en/publications-data/ download-todays-data-geographic-distribution-covid-19-cases-worldwide ECDC) on 11th May 2020. We took a subset of the case fatalities for 26th April 2020 corresponding to NextStrain's sequences. The region of North America includes the following countries: Canada, Mexico, Panama, USA. The region of South America includes the following countries: Brazil, Chile, Colombia, Ecuador, Peru, Uruguay. The 5 deaths from Georgia were subtracted from Europe and added to Asia, because Georgia is classified as Asia in the NextStrain data. In addition, there are 7 deaths not classified in any of the regions by ECDC. These are labelled as "Cases on an international conveyance Japan" and seem to correspond to deaths on cruise ships. We excluded these completely. The air passenger data have been obtained through the commercial provider SABRE [18]. Data are consolidated for the years 2019 and 2020.

References

1. Akaike, H.: Information theory and an extension of the maximum likelihood principle. In: Parzen, E., Tanabe, K., Kitagawa, G. (eds.) Selected Papers of Hirotugu Akaike. Springer Series in Statistics, pp. 199–213. Springer, New York (1998). https://doi.org/10.1007/978-1-4612-1694-0_15
2. Beaulieu, J.M., O'Meara, B.C.: Detecting hidden diversification shifts in models of trait-dependent speciation and extinction. Syst. Biol. **65**(4), 583–601 (2016). https://doi.org/10.1093/sysbio/syw022
3. Cole, D.J.: Parameter redundancy and identifiability in hidden Markov models. METRON **77**, 105–118 (2019). https://doi.org/10.1007/s40300-019-00156-3
4. FitzJohn, R.G.: Diversitree: comparative phylogenetic analyses of diversification in R. Methods Ecol. Evol. **3**, 1084–1092 (2012). https://doi.org/10.1111/j.2041-210X.2012.00234.x
5. Gelman, A., Carlin, J.B., Stern, H.S., Rubin, D.B.: Bayesian Data Analysis, 2nd edn, pp. 296–297. CRC Press, Boca Raton (2004)
6. Geoghegan, J.L., et al.: Genomic epidemiology reveals transmission patterns and dynamics of SARS-CoV-2 in Aotearoa New Zealand. Nature Commun. **11**(1), 6351 (2020). https://doi.org/10.1038/s41467-020-20235-8, https://www.nature. com/articles/s41467-020-20235-8
7. Hadfield, J., et al.: Nextstrain: real-time tracking of pathogen evolution. Bioinformatics **34**(23), 4121–4123 (2018). https://doi.org/10.1093/bioinformatics/bty407

8. Höhna, S., et al.: RevBayes: Bayesian phylogenetic inference using graphical models and an interactive model-specification language. Syst. Biol. **65**(4), 726–736 (2016). https://doi.org/10.1093/sysbio/syw021

9. Kermack, W.O., McKendrick, A.G., Walker, G.T.: A contribution to the mathematical theory of epidemics. Proc. Roy. Soc. Lond. Ser. A Containing Papers Math. Phys. Character **115**(772), 700–721 (1927). https://doi.org/10.1098/rspa.1927.0118

10. Lemieux, J.E., et al.: Phylogenetic analysis of SARS-CoV-2 in Boston highlights the impact of superspreading events. Science **371**(6529) (2021). https://doi.org/10.1126/science.abe3261, https://science.sciencemag.org/content/371/6529/eabe3261

11. Newton, M.A., Raftery, A.E.: Approximate Bayesian inference with the weighted likelihood bootstrap. J. Roy. Stat. Soc. Ser. B (Methodol.) **56**(1), 3–26 (1994)

12. Pan, B., et al.: Identification of epidemiological traits by analysis of SARS-CoV-2 sequences. Viruses **13**(5), 764 (2021). https://doi.org/10.3390/v13050764

13. Paradis, E., Schliep, K.: ape 5.0: an environment for modern phylogenetics and evolutionary analyses in R. Bioinformatics **35**, 526–528 (2019)

14. Popa, A., et al.: Genomic epidemiology of superspreading events in Austria reveals mutational dynamics and transmission properties of SARS-CoV-2. Sci. Transl. Med. **12**(573) (2020). https://doi.org/10.1126/scitranslmed.abe2555, https://stm.sciencemag.org/content/12/573/eabe2555

15. Price, M.N., Dehal, P.S., Arkini, A.P.: Fasttree: computing large minimum evolution trees with profiles instead of a distance matrix. Mol. Biol. Evol. **26**(7), 1641–1650 (2009). https://doi.org/10.1093/molbev/msp077

16. R Core Team: R: A Language and Environment for Statistical Computing. R Foundation for Statistical Computing, Vienna, Austria (2020). https://www.R-project.org/

17. Revell, L.J.: phytools: An R package for phylogenetic comparative biology (and other things). Methods Ecol. Evol. **3**, 217–223 (2012)

18. SABRE: Sabre market intelligence platform (2020). https://www.sabreairlinesolutions.com/images/uploads/AirVision-Market-Intelligence_GDD_Profile_Sabre.pdf

19. Sagulenko, P., Puller, V., Neher, R.A.: TreeTime: maximum-likelihood phylodynamic analysis. Virus Evol. **4**(1), vex042 (2018). https://doi.org/10.1093/ve/vex042

20. Schwarz, G.E.: Estimating the dimension of a model. Ann. Stat. **6**(2), 461–464 (1978). https://doi.org/10.1214/aos/1176344136

21. Sjaarda, C.P., et al.: Phylogenomics reveals viral sources, transmission, and potential superinfection in early-stage COVID-19 patients in Ontario, Canada. Sci. Rep. **11**(1) (2021). https://doi.org/10.1038/s41598-021-83355-1, https://www.nature.com/articles/s41598-021-83355-1

22. Takahashi, S., Greenhouse, B., Rodríguez-Barraquer, I.: Are seroprevalence estimates for severe acute respiratory syndrome coronavirus 2 biased? J. Infect. Dis. **222**(11), 1772–1775 (2020). https://doi.org/10.1093/infdis/jiaa523

23. Yanev, N.M., Stoimenova, V.K., Atanasov, D.V.: Branching stochastic processes as models of Covid-19 epidemic development. arXiv e-prints (2020)

Explainable AI Models for COVID-19 Diagnosis Using CT-Scan Images and Clinical Data

Aicha Boutorh[1,2](✉)(iD), Hala Rahim[2], and Yassmine Bendoumia[2]

[1] Laboratory for Research in Artificial Intelligence (LRIA),
Department of Computer Science, USTHB, Algiers, Algeria
`aboutorh@usthb.dz`
[2] Department of Computer Science, Faculty of Sciences, Algiers 1 University -
Benyoucef Benkhedda, Alger Ctre, Algeria

Abstract. The pandemic of COVID-19 has had a significant impact on global health and is becoming a major international concern. Fortunately, early detection helped decrease its number of deaths. Artificial Intelligence (AI) and Machine Learning (ML) techniques are a new era, where the main objective is no longer to assist experts in decision-making but to improve and increase their capabilities and this is where interpretability comes in. This study aims to address one of the biggest hurdles that AI faces today which is public trust and acceptance due to its black-box strategy. In this paper, we use a deep Convolutional Neural Network (CNN) on chest computed tomography (CT) image data and Support Vector Machine (SVM) and Random Forest (RF) on clinical symptoms data (Bio-data) to diagnose patients positive for COVID-19. Our objective is to present an Explainable AI (XAI) models by using the Local Interpretable Model-agnostic Explanations (LIME) technique to identify positive patients to the virus in an interpreted way. The results are promising and outperformed the state of the art. The CNN model reached an Accuracy and F1-Score of 96% on CT-scan images, and SVM outperformed RF with Accuracy of 90% and Specificity of 91% on Bio-data. The interpretable results of XAI-Img-Model and XAI-Bio-Model, show that LIME explanations help to understand how SVM and CNN black box models behave in making their decision after being trained on different types of COVID-19 dataset. This can significantly increase trust and help experts understand and learn new patterns for the current pandemic.

Keywords: Explainable AI · Deep learning · CNN · Black box classifiers · LIME · COVID-19 diagnosis · Image data · Clinical data

1 Scientific Background

In December 2019, a novel virus named COVID-19 emerged in the city of Wuhan, China. It caused widespread infections and deaths due to its contagious characteristics. COVID-19 is an infectious disease caused by Severe acute respiratory

D. Chicco et al. (Eds.): CIBB 2021, LNBI 13483, pp. 185–199, 2022.
https://doi.org/10.1007/978-3-031-20837-9_15

syndrome Coronavirus 2 (SARS-CoV-2). It was declared pandemic by the World Health Organization (WHO) on March 11, 2020. The virus is so perilous and can provoke the death of people with weakened immune systems. The global pandemic has motivated the research community to come with cutting edge research for combating this virus [1,2].

Various methods have been used to diagnose COVID-19, containing blood tests, PCR and a variety of medical imaging techniques. The main digital support that have tackled the pandemic with novel methods comes from the Artificial Intelligence (AI) and Machine Learning (ML) community in the form of automated COVID-19 detection. AI techniques were used for different types of data including non-imaging and imaging datasets [3]. Two medical imaging techniques, X-ray and CT-scan, are employed to diagnose COVID-19 using AI and ML methods. The most AI algorithms used for different types of data, including non-imaging and imaging datasets [3], are Convolutional Neural Network (CNN), Support Vector Machines (SVM), Logistic Regression, Decision Trees (DT) and Random Forest (RF) [4].

De Moraes et al. [5] used SVM and data from emergency care admission exams (RT-PCR) to detect COVID-19 cases. They collected data from 235 patients of which 43% were confirmed COVID-19 cases. They trained five machine learning algorithms and found out that the SVM had the best performance with an accuracy of 85%. The authors concluded that the method could be used to target which patient needs a laboratory COVID-19 tests done on them. Mei et al. in [6] is one of the contributions that used AI models to generate the probability of a patient being positive for COVID-19. The authors applied a CNN on a chest CT scan that reported a sensitivity of 83.6% and a specificity of 75.9%, and SVM on clinical information with 80.6% and 68.3% respectively. Wang et al. [7] used CNN with a dataset comprising of 13,800 chest X-ray radiography images from 13,725 patients so as to try and provide clinicians with a deeper insight into the critical factors affecting with COVID-19 cases. They reported an accuracy, sensitivity and positive prediction value (PPV) of 92.6%, 87.1% and 96.4% respectively. Li et al. [8] employed CNN for the detection of COVID-19. The authors extracted visual features from volumetric chest CT images of COVID patients and classified them. They reported that the method was not only able to detect COVID-19 case but also to distinguish it from other community acquired pneumonia and non-pneumonic lung diseases. The authors concluded that CNN with X-ray imaging might extract significant biomarkers related to COVID-19.

Deep Neural Networks (DNN) can perform wonderful feats, thanks to their extremely large and complicated web of parameters, but their complexity is also their curs. The inner workings of NNs are often a mystery even to their creators. This is a challenge that has been troubling the AI community since deep learning started to become popular in the early 2010s. Research showed that the use of black-box models such as SVM or Neural Nets (NN) was very high followed by Decision Trees (DT), which are explainable models and highlighted that the use of NN has decreased over time which is, very probably, due to their inexplicable behavior [9].

So, one of the biggest hurdles that AI faces today is public trust and acceptance. People struggle to trust the decisions and answers that AI-powered tools provide. AI does not explicitly share how and why it reaches its conclusions. Interpretability is one of the most common reasons limiting the black-box models to be accepted and used in critical domains such as the medical one. Explainable Artificial Intelligence (XAI) is a set of processes and methods that allow human users to understand and trust the results and conclusions created by machine learning algorithms. Demanding more interpretability and explainability by the healthcare community creates, indeed, more challenges for the AI and ML technologies [9].

Among the developed explainable methods comes LIME (Model-agnostic Explanations) [10] which is a local interpretation technique that aims to explain the conditional interaction between the decisions and the attributes concerning a single prediction. LIME explains the predictions of any classifier by treating the model as a black box and learning an interpretable model locally around a prediction. LIME was applied to different disease diagnosis systems in healthcare domain such as Chronic Wound classification [11], Glioblastoma Multiforme (GBM) diagnosis [12], Lung cancer Diagnosis [13], Parkinson's Disease (PD) diagnosis [14] and others. Several research works have been done on AI/ML models' interpretation [9]. "Why should I trust it?" is the most powerful question users ask when it comes to using a deployed model to make decisions. ML explainability has the power to break even the models with the highest accuracy.

Our goal in this research is to give an interpretation of the COVID-19 cases separation generated by our developed models that could perform multi-classification task using different types of datasets. The objectives of our study in this paper are:

- Build a powerful CNN COVID-19 diagnosis model using deep learning and transfer learning methods for CT-Scan imaging data.
- Build a high-performance COVID-19 diagnostic model on a non-imaging dataset, including clinical data, by comparing multiple ML classifiers like SVM and RF.
- Create an Explainable AI (XAI) system by applying interpretable method as LIME to the developed classifiers in order to convert complex black-box AI models to more understandable glass box. The aim is to give the user the ability to follow the reasoning behind the AI decision.

The rest of the paper is organized as follow: Sect. 2 describes the datasets used in this study as well as the design of CNN deep learning, SVM and RF models trained for the COVID-19 detection. Section 3 presents the experimental results of this empirical evaluation, the comparison of our models to the literature in addition to the LIME interpretability results for the clinical and imaging data. We conclude the paper in Sect. 4 and give some future work.

2   Materials and Methods

In this study we propose an Explainable AI diagnosis model for multiple databases: Clinical data (laboratory test results, reported symptoms, history of exposure etc.) as well as chest imaging findings for rapid identification of patients with COVID-19. Our main objective is to interpret the ML black boxes to explain the cases of the diagnosis decision for the human experts.

2.1   Datasets Description and Preprocessing

Data is an essential element for the efficient implementation of ML methods. The combination of AI and data sets generates a practical COVID-19 diagnosis solution [1]. In this work we are interested in the dataset that contains the information of clinical data and chest imaging findings, so that we could carry out our study on multi-datasets [6]. The data was acquired from 905 patients for whom there was a clinical concern of COVID-19. It was collected between 17 January and 3 March 2020 from 18 medical centers in 13 provinces in China. The final dataset was a cleaned and filtered version (3 classes) of the initial dataset (12 classes).

Image Dataset: The first dataset used in our study represents the chest CT scans [6]. Each image is accompanied by a set of 26 attributes such as patient ID, age, date and location, and included 488 men and 417 women. Each patient may have multiple images taken on different days. The original images dataset is composed of 12 classes including tuberculosis, viral, SARS, etc. The CT scan dataset used for this study consisted of 619 samples of which 520 samples belong to the COVID-19 class, 81 samples belong to the pneumonia class, and 18 samples belong to the normal class. The other classes were eliminated for irrelevance. ImageDataGenerator was used to preprocess image batches before training: resizing the image dataset to 224*224*3 pixels, thresholding the image to remove any very bright pixels and repair missing areas, and normalizing the pictures by scaling them so their pixel values within the run [0, 1].

Bio-Dataset: The second dataset represents the clinical information named as Bio-Dataset [6]. The data contains 619 rows that represent patients with 23 features: patient-ID, RT-PCR, went-ICU, hypertension, extubated, lymphocyte-count, neutrophil-count, leukocyte-count, age, temperature, pO2-saturation, sex, heart-disease, intubation-present, offset, diabetes, cancer, fatigue, in-ICU, intubated, needed supplemental O2, white blood cell count and survival. Patients are classified into three classes: COVID-19, Pneumonia and Normal. To deal with the missing values in Bio-dataset, we have filling them using the mean methods. The categorical features were transformed to numerical values by using encoding method. For clinical notes feature (text type) we only put in consideration important notes: hypertension, diabetes, coronary heart, cancer and fatigue, we then transformed them into Boolean attributes mentioning 1 if it is Yes, and 0 if it is not.

2.2   Models Design

The objective of our study is to implement an explainable AI system on different types of data that will allow to detect COVID-19. For this end, we created two AI/ML models to detect a patient being COVID-19. The first model is a deep learning Artificial Neural Network model with Convolutional Neural Network (CNN or ConvNet) architecture based on a chest CT scan data, and the second model is supervised Support Vector Machine (SVM) and Random Forest (RF) classifiers that were trained and compared on clinical input data.

CNN Architecture for Image Data: For the classification of chest CT scan data, we used Convolutional Neural Network (CNN) which is a specific kind of deep learning architecture. One of the great advantages of CNN is its ability to auto-extract the features it will need with consideration of the spatial structure of the image. In the proposed architectures, VGG-16, ResNet-50, and InceptionV3 were employed as convolutional bases of the model. All the CNNs were pre-trained on ImageNet.

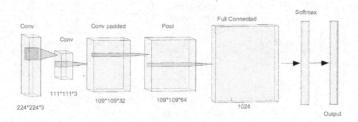

Fig. 1. CNN network design of inception-V3 used model.

Our CNN deep-learning model was developed using transfer learning using of certainly considered one among 3 ultra-modern pre-trained model (Inception-V3). The implemented transfer learning models are divided into two parts Convolutional base and Classifier. The convolutional base is used as a spatial feature extractor and the classifier predicts the class label based on features extracted by CNN. InceptionV3 is a pre-trained model that belongs to transfer learning models (Inception family), widely used for image recognition models. The classifier, as shown in Fig. 1, consists of a global average and max pooling and a fully connected layer with the ReLU activation function and Softmax function for the label prediction. This model is also composed of several convolutional layers, pooling layers in parallel. An Adam optimizer was used to optimize the weights of the models. The learning rate was set to 0.001 and batch size of 32. To fight against overfitting, we used more than one technique: Dropout regularization with a rate of 0.1 and data augmentation which consists of randomly rotating the image and zooming with a range of 0.2.

Classifier Model for Bio-Data: Regarding our second model, we have used supervised classifier, where the algorithm is trained on clinical input data that has been labeled for a particular output. We have applied two Machine Learning models SVM and RF. The Fig. 2 represents the phases of realization of our Bio-Classifier. The model is composed of three steps. The first step is collecting and preprocessing data by cleaning and selecting important attributes. The second step is the classification algorithms, we applied two algorithms SVM and RF to test the performance of each and choose the best model. And the third step, testing and evaluating the modes for performance comparison. After the performance evaluation, SVM was reported as the best model for the Bio-dataset.

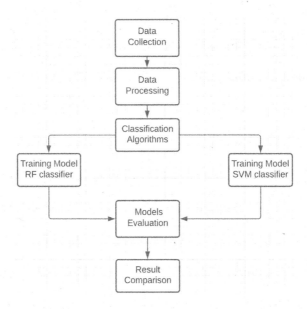

Fig. 2. Bio-data classifier architecture.

2.3   Explainability and Interpretability

We applied two AI/ML models to detect a patient being COVID-19. We named the first model XAI-Img-model, which is based on a chest CT scan where we used a deep learning CNN architecture. The second was named XAI-Bio-model, and it is based on clinical information where we have applied SVM method after showing its performance comparing to RF. For both models, we have interpreted their results using the LIME method as shown this the Fig. 3. As presented earlier, LIME uses the linear model to determine which features were most contributory to the model's prediction for each learnt example.

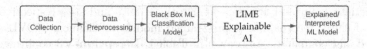

Fig. 3. Explainable/interpretable model design using LIME

XAI-Img-Model: As transparency, interpretability, and explainability are necessary in radiology to build patient and provider trust [15], we are proposing to employ the LIME algorithm to generate explanations for the predictions of the best deep learning CNN classifier architecture we have trained on imaging dataset. By applying LIME to our CNN trained model, we can conduct informed feature engineering based on clearly inconsequential features that we see for image data. LIME considers features to be super pixels of the image.

XAI-Bio-Model: For patient's detection, in case of COVID-19, it is necessary to understand the symptoms to be able to decide if the case is positive or negative. LIME was used to interpret our SVM model decision. This interpretable technique has made the black-box model to be of high Interpretability. This phase aimed to interpret the selected model locally and define how each feature affects the final decision and on which features the model relies on the most.

3   Results

Our experimental study is based on 2 types of data; each data is composed of 619 samples and three classes: Normal, COVID-19 and Pneumonia. The models were evaluated and compared to each other as well as to the state of the art [6] that uses the same dataset. The best model is not necessarily the one with higher Accuracy, therefore, multiple performance classification measures were used to select the best model/architecture. The best model for each type of data is then interpreted using the LIME method. The experimental results as well as the prediction explanation are given in the following subsections.

3.1   Deep CNN for Image-Data Experimentation Results

In our first series of experiments we employed three kinds of CNNs models on image dataset and fine-tuned them in keeping with our requirements. Our model was developed with the aid of using transfer learning using of certainly considered one among 3 ultra-modern pre-trained models (VGG16, ResNet-50, and Inception-v3) as a backbone. These 3 networks have three fully connected layers (FC), the last one is used for classification purpose. We initialized the last FC layer according to the target dataset. An Adam optimizer was used to optimize the weights of the models. The learning rate was set to 0.001 and batch size of 32. We trained the 3 models for several epochs. The CT scan images set was split into train and test sets at a ratio of 80% and 20% respectively. For the set of check records, the 3 pre-trained models to diagnose COVID-19 are evaluated and compared. As proven in Table 1, **Inception-V3** confirmed great diagnostic performance. Our deep CNN model outperformed the alternative pre-trained fashions within the testing dataset, regarding the **Accuracy** and **F1-Score** of **96%**. The same value obtained from these metrics can be interpreted mainly by the fact that we are dealing with a multi-class classification, where each test data must belong to a single class and not to multiple labels. The Specificity and Sensitivity are both hight for all the models, they reached values of 94% and 80% respectively.

Table 1. Deep CNN Classification Performance obtained by Inceptionv3, VGG16 and Resnet50 for Image-Dataset. Evaluation Metrics % used: Acc: Accuracy, Spec: Specificity, Sens: Sensitivity and F1-Score

Pre-trained model	Epochs	Loss	Acc	Spec	Sens	F1-Score
InceptionV3	20	0.04	**0.96**	0.94	0.80	**0.96**
VGG16	20	0.31	0.87	0.94	0.80	0.87
Resnet-50	100	0.55	0.83	0.94	0.80	0.83

3.2   Classifiers for Bio-Data Experimentation Results

In the second series of experiments, we have applied two supervised learning algorithms on Bio-data, which are Random Forest (RF) and Support Vector Machine (SVM). Each of the two algorithms needs different inputs parameters to increase the predictive power of the model. The Bio dataset was split into Train, Validation and Test sets at a ratio of 6:2:2 which means 60% training set, 20% validation set and 20% test set. To find an optimal hyper-parameter we can just try all combinations and see what parameters work best. The idea is to create a grid of hyper-parameters and just try all of their combinations.

RF: There are fundamentally 2 elements for RF which can be tuned to work on the prescient force of the model: Features which improve forecasts of the model like n-estimators which present the quantity of trees you need to fabricate and max-features present greatest number of elements Random Forest is permitted to attempt in individual tree. For the hyper-parameters of RF, we tried different values. Table 2 shows the trained and the best parameters for RF classifier. The best parameters reached an Accuracy of 83% on the training set and 90% on the validation set.

Table 2. Trained Parameters and Best Parameters for RF Classifier Model. The reached **Accuracy** for the best parameters was **83% for the Training set** and 90% **for the Validation set.**

Grid parameters	Trained values	Best values
Criterion	[gini, entropy]	Gini
Max_depth	[4, 5, 6, 7, 8, 10]	7
Max_features	[auto, sqrt, log2]	Auto
N_estimators	[21, 30, 60, 100, 200, 500]	100

SVM: The SVM algorithm was used on Bio-dataset as a second classifier. We have trained different parameters in order to increase the predictive power of the model. SVM has a set of parameters as the regularization parameter C and the kernel (similarity function) type to be used in the algorithm. SVM, was evaluated to predict COVID-19 using clinical information and after various emphases, we arrived at the best values of C=1, Gama = 0.1 and Kernel = rbf. Table 3 shows the trained and the best parameters for SVM classifier. The best parameters reached an Accuracy of 92% on both the training and validation sets.

Table 3. Trained parameters and best parameters for SVM classifier model. The reached **Accuracy** for the best parameters was **92% for the Training set** and **92% for the Validation set.**

Grid parameters	Trained values	Best values
C	[1, 10, 30, 60, 80, 100, 200,400, 600, 1000]	1
Gama	1, 0.1, 0.01, 0.001, 0.0001]	0.1
Kernel	[linear, rbf]	rbf

SVM and RF Comparison:
Comparing the two created classifier models, it is clear that the model based RF is as performant as the model based on SVM. By analyzing the results presented in Table 4 we can conclude that the SVM model slightly outperformed RF in its abilities of 70% of Sensitivity and 90% of Accuracy and F1-Score. Both models reached high Specificity of up 90%.

Table 4. Performance comparison of RF and SVM classification models.

Model	Accuracy	Specificity	Sensitivity	F1-Score
RF	0.85	0.90	0.60	0.85
SVM	**0.90**	**0.91**	**0.70**	**0.90**

From the ROC Curve presented in Fig. 4, it is evident that the AUC for the SVM ROC curve is higher than that for the RF ROC curve. Both SVM and RF perform well on the dataset, but we are choosing SVM since ROC curve is overall higher.

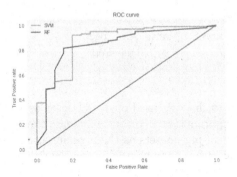

Fig. 4. ROC curve for SVM and RF

3.3   Comparison Study

After various tests, we have come to the resolution that the best model that fits well our CT scan dataset is the deep CNN InceptionV3 with 96% Accuracy, 80% Sensitivity and 94% Specificity. Compared to the state of the art, authors in [6] used CNN architecture for the same image dataset and reported 75.9% specificity and 83.6% sensitivity. They applied SVM, RF and MLP on Bio dataset. The results showed that the best model that fit well on this dataset was MLP with specificity 68.3%. On the other side, we applied on the same clinical dataset our SVM classifier that outperformed MLP with a specificity of 91%, the reason can be the lack of data which can cause overfitting in complicated models. The comparison study is presented in Table 5. Our ML models had a better specificity and a close sensitivity. By analyzing the results, we can conclude that our deep CNN and SVM models are most appropriate for this types of datasets with high performance.

Table 5. Comparison between our models and the models in [6]. **CNN-CTscan** : our deep CNN model on Image data. **SVM-Bio**: our SVM model on Clinical data.

Metrics	CNN-CTscan	CNN [6]	SVM-Bio	MLP [6]
Specificity	**94%**	75.9%	**91%**	68%
Sensitivity	**80%**	84%	**70 %**	81%

3.4   Explainability/Interpretability Results

Interpretability aims increasing model trustworthiness especially in critical domains such as the medical one. In this study, LIME was used ·for the local interpretability of the best CNN and SVM models applied on COVID-19 multi-datasets. In this section, we present the explanation of individual prediction for three instances of different classes so the decider can understand why the model predicted a specific class for a particular test instance.

XAI-Img-Model: In order to explain the way our CNN model works on image data, we decided to interpret the deep black-box model using LIME as shown in Fig. 5, which is a plot explanations of the deep learning CNN model on image data to interpret the decision and understand how it was made. This explainable model is named "XAI-Img-model".

Super pixels hued green as shown in Fig. 5 demonstrate areas that were generally contributory toward the anticipated class. Alternately, super pixels shaded red show districts that were generally contributory against the anticipated class. For the class predicted COVID-19, as shown in Explanation (A) of Fig. 5 with 97%, it is predicted positively at the level of the shoulder, and negatively inside the lungs and the neck. The class predicted Normal in the Explanation (B) with 97.5% predicting positively inside the lungs, and negatively inside the lungs, the neck and at the level of the shoulder. For the class predicted Pneumonia, Explanation (C), with 99.1% predicting positively inside the lungs and at the level of the shoulder, and negatively on the side of thoracic cavity.

XAI-Bio-Model: For this model, we aim to explain the way our SVM works on clinical data by applying the LIME interpretation. This explainable model is named "XAI-Bio-model". A set of three instances was chosen from the test set, then the explanation was plotted to interpret the black-box model and understand how our SVM uses the features to make its decision. The interpretable results are shown in Fig. 6. The information given by LIME represent: the prediction probability of the three classes (COVID-19, Normal, Pneumonia) given in the leftmost box, the middle chart shows the important features with their bounding values and the table below is the actual values of the corresponding features in the observation row passed as input.

We choose randomly one individual to visually show the effects of the features on its class. As we are able to see in the explanation (A) of the Fig. 6, the LIME results indicates that our model predicted a COVID-19 with a probability of 1.00, this means that the possibility that this selected person has a COVID-19 disease is 100%, 0% for Normal and 0% for Pneumonia. In this explication we notice how the needed_supplemental_O2 and intubation_present switched the decision towards COVID-19 class. Also, we can see that the effect of the sex being male has a much worse effect as being in 0.09 bucket. The model's decision was affected more by the first three features. For the explanation (B), the results indicate that our model predicts a probability that this selected person is Normal 71%, 29%

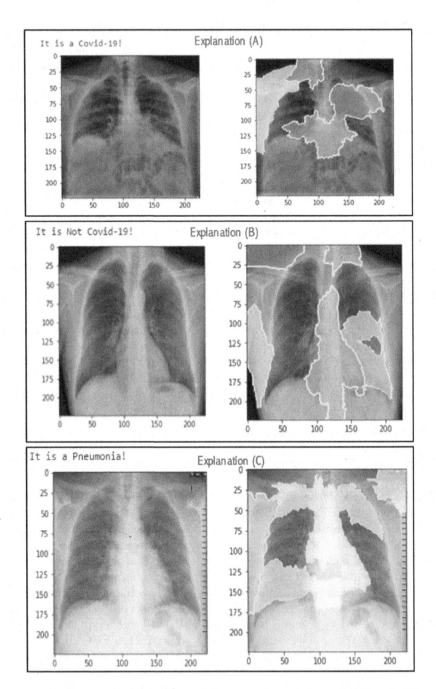

Fig. 5. LIME explanation of three test instances for our deep CNN based CT-scan COVID-19 Dataset: "XAI-IMG-Model": (A) Class COVID-19; (B) Class Normal; (C) Class Pneumonia.

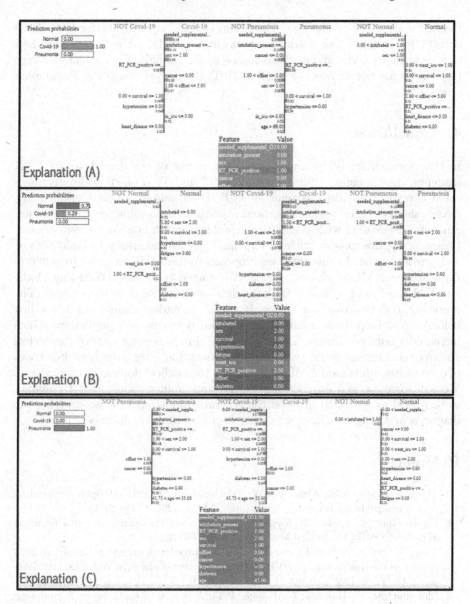

Fig. 6. LIME Explanation of three test instances for our SVM based Demographic/Clinical COVID-19 Data: "XAI-BIO-Model": (A) Class COVID-19; (B) Class Normal; (C) Class Pneunomomia.

COVID-19 and 0% Pneumonia. We can see the effect of needed-supplemental-O2 of 0.15 and intubation present of 0.09 on the class percentage of COVID-19 for this patient. The model decided that the person is normal based on the three last features for NOT COVID-19 and the three first ones for NOT Pneumonia. In the explanation (C) results show Pneumonia disease is 100%, 0% for COVID-19 and 0% for Normal. The needed_supplemental and the intubation_present had a huge impact on classifying the instance as Pneumonia. Five attributes were responsible for the decision of NOT COVID-19 and privileged the Pneumonia class.

4   Conclusion

In this research study, we investigated the power of the deep learning CNN model on chest imaging findings as well as RF and SVM on clinical data which could effectively predict outcome for COVID-19 patients. The developed ML models showed a good performance and reached a hight evaluation metric values. Since we are dealing with a healthcare problem which is a very sensitive field, transparency, interpretability, and explainability are necessary to build patient and provider trust. In this work we proposed to use LIME in order to generate explication for ML black-box models. We named our models XAI-Img-Model and XAI-Bio-Model on image CT-scan data and Clinical data respectively. The proposed models proved to be highly useful in understanding the black box behavior and help domain expert understand decisions and predictions which increase its trustworthiness. The study is still in its earlier stages of theoretical research. As ongoing work, using other datasets and integrating multiple types of data for the same model will be helpful as the medical domain is very critical. We will focus as well on checking the interpretability using other techniques wether local or global to increase the trustworthiness, as well as, studying the diagnosis and prognosis interpretability for other complex diseases.

References

1. Shuja, J., Alanazi, E., Alasmary, W., Alashaikh, A.: COVID-19 open source data sets: a comprehensive survey. Appl. Intell. **51**(3), 1296–1325 (2021)
2. Li, J., Guo, X.: Global deployment mappings and challenges of contact-tracing apps for COVID-19. Available at SSRN 3609516 (2020)
3. Yang, W., et al.: Clinical characteristics and imaging manifestations of the 2019 novel coronavirus disease (COVID-19): a multi-center study in Wenzhou city, Zhejiang, China. J. Infection **80**(4), 388–393 (2020)
4. Mohamadou, Y., Halidou, A., Kapen, P.T.: A review of mathematical modeling, artificial intelligence and datasets used in the study, prediction and management of COVID-19. Appl. Intell. **50**(11), 3913–3925 (2020)
5. Alballa, N., Al-Turaiki, I.: Machine learning approaches in COVID-19 diagnosis, mortality, and severity risk prediction: a review. Inform. Med. Unlocked **24**, 100564 (2021)
6. Mei, X., et al.: Artificial intelligence-enabled rapid diagnosis of patients with COVID-19. Nat. Med. **26**(8), 1224–1228 (2020)
7. Islam, M.Z., Islam, M.M., Asraf, A.: A combined deep CNN-LSTM network for the detection of novel coronavirus (COVID-19) using X-ray images. Inform. Med. Unlocked **20**, 100412 (2020)

8. Ismael, A.M., Şengür, A.: Deep learning approaches for COVID-19 detection based on chest X-ray images. Expert Syst. Appl. **164**, 114054 (2021)

9. Ahmad, M.A., Eckert, C., Teredesai, A.: Interpretable machine learning in health-care. In: Proceedings of the 2018 ACM International Conference on Bioinformatics, Computational Biology, and Health Informatics, pp. 559–560, August 2018

10. Ribeiro, M.T., Singh, S., Guestrin, C.: Why should i trust you? Explaining the predictions of any classifier. In: Proceedings of the 22nd ACM SIGKDD International Conference on Knowledge Discovery and Data Mining, pp. 1135–1144, August 2016

11. Sarp, S., Kuzlu, M., Wilson, E., Cali, U., Guler, O.: The enlightening role of explainable artificial intelligence in chronic wound classification. Electronics **10**(12), 1406 (2021)

12. Rucco, M., Viticchi, G., Falsetti, L.: Towards personalized diagnosis of glioblastoma in fluid-attenuated inversion recovery (FLAIR) by topological interpretable machine learning. Mathematics **8**(5), 770 (2020)

13. Meldo, A., Utkin, L., Kovalev, M., Kasimov, E.: The natural language explanation algorithms for the lung cancer computer-aided diagnosis system. Artif. Intell. Med. **108**, 101952 (2020)

14. Magesh, P.R., Myloth, R.D., Tom, R.J.: An explainable machine learning model for early detection of Parkinson's disease using LIME on DaTSCAN imagery. Comput. Biol. Med. **126**, 104041 (2020)

15. Geis, J.R., et al.: Ethics of artificial intelligence in radiology: summary of the joint European and North American multisociety statement. Can. Assoc. Radiol. J. **70**(4), 329–334 (2019)

The Need of Standardised Metadata to Encode Causal Relationships: Towards Safer Data-Driven Machine Learning Biological Solutions

Beatriz Garcia Santa Cruz[1]([✉])(iD), Carlos Vega[2](iD), and Frank Hertel[1]

[1] National Department of Neurosurgery, Centre Hospitalier de Luxembourg,
Luxembourg City, Luxembourg
garciasantacruz.beatriz@gmail.com
[2] Luxembourg Centre for Systems Biomedicine, University of Luxembourg,
Esch-sur-Alzette, Luxembourg
carlos.vega@uni.lu

Abstract. In this paper, we discuss the importance of considering causal relations in the development of machine learning solutions to prevent factors hampering the robustness and generalisation capacity of the models, such as induced biases. This issue often arises when the algorithm decision is affected by confounding factors. In this work, we argue that the integration of research assumptions as causal relationships can help identify potential confounders. Together with metadata information, it can enable meta-comparison of data acquisition pipelines. We call for standardised meta-information practices as a crucial step for proper machine learning solutions development, validation, and data sharing. Such practices include detailing the data acquisition process, aiming for automatic integration of causal relationships and actionable metadata.

Keywords: Confounders · Causality · Metadata · Machine learning · Systems biology

1 Introduction

The number of scientific publications in the biological field employing machine learning (ML) is rapidly growing [1]. Both as a result of better access to larger amounts of data generated using the latest technology (e.g., high throughput screening) and the computational capacity together with the fast development in the ML area, especially in deep learning (DL) with the use of convolutional neural networks (CNN) and generative adversarial networks (GAN) [2,3].

Biological systems are complex, though the advent of ML has shown a promising approach to working with data stemming from such complex phenomena for conducting data analysis and prediction. However, such progress has not come without challenges, e.g., model interpretation of ML and DL models is a usual hurdle but also a common neglected requirement. The heterogeneity

D. Chicco et al. (Eds.): CIBB 2021, LNBI 13483, pp. 200–216, 2022.
https://doi.org/10.1007/978-3-031-20837-9_16

and multi-modal nature of biological data also presents challenges such as the curse of dimensionality, which may lead to data sparsity or multicollinearity [4]. Multi-modal and heterogeneous data also call for better metadata to convey the acquisition details and facilitate the comparison and contrast of datasets and acquisition pipelines.

Such trend and their direct consequences in biological healthcare applications call for standardised guidelines to ensure the quality of each stage of the research and application pipelines. This is specially important when comparing data acquisition pipelines, mixing datasets or comparing ML performance across datasets (e.g., for external validation) which may potentially stem from different data generation processes.

Ensuring good quality during the whole process, from data generation to model deployment, is a complex and ambitious task, still fundamental to building correct ML models to study biological phenomena. The most notable approaches include some of the current guidelines.. Among other objectives, these guidelines aim to establish better data sharing and appropriate foundations for good appraisal and reproducibility. The data sharing goal aims to ensure good data management not only to advance in knowledge discovery and innovation but also to allow for proper data reuse. Better appraisal and reproducibility can be achieved through standardised reporting guidelines that guarantee the reporting of key dataset elements (for example dataset generation details) as an essential step for dataset comparison and validation. In this way, remarkable efforts have been done in the recent years. Here we highlight the FAIR principles [5] and the DOME recommendations [6].

The FAIR principles (Findability, Accessibility, Interoperability, and Reusability) aim to increase data usability, with special emphasis on machine-readable and actionable datasets. This need arises because machines, in contrast to humans, lack a natural ability to identify and interpret the context, becoming more likely to make errors contextualising data. However, machines can overcome humans' main limitations operating at the scope, scale, and speed that the current e-Science scenario requests. Thus, different mechanisms and protocols seeking machine self-guidance for data exploration need to be developed [5].

More specific to the area of applied ML for biological analysis, we can find DOME (Data, Optimisation, Model and Evaluation), a community-wide collection of recommendations focused on standardised review guidelines for proper reporting of supervised ML in biological studies. DOME's impact is not limited to the evaluation of publications' results individually, but it opens the door to better meta-analysis of ML datasets which enables comparison of methods and avoidance of unnecessary repetition of data generation to answer to new research questions [6]. However, today such meta-analysis is not easily feasible due to the lack of data sharing standardisation.

The rest of the paper is structured as follows: with FAIR and DOME as guidance, we discuss the importance of considering potential confounders in the data, especially after the paradigm change from classical statistical modelling to ML. Below, we explore its impact on the ML-based biological applications. Finally, we discuss potential solutions with a special focus on the standardised metadata that aims to encode causal relationships. Although every step is critical to develop better models, this paper focuses on the datasets and their standardised reporting in the context of biological research.

2   Considerations for the Development and Reporting of ML Solutions

The study of biological systems involves either inferences or predictions. Inference aims to create a mathematical model about the data-generation process, testing a hypothesis or formalising our understanding of how the studied systems behave. It is used to understand the mechanism of the studied event, e.g., how the accumulation of one specific protein affects the system. In contrast, the purpose of prediction is to forecast future behaviour, without necessarily understanding the mechanism behind it, e.g., to predict which treatment is better based on the specific level of a determined protein. Despite the fact that both statistics and ML can be used to predict and make inferences, traditionally, statistical methods have been applied for inference whilst ML methods were employed for prediction [7]. The choice between prediction or inference depends on the ultimate analysis goal.

In short, the statistical approaches are useful when we want to understand the influence of each variable, but in general (not always), have less predictive power, frequently because only few variables and linear relationships are considered. Conversely, when large and high dimensional datasets are analysed with prediction as a goal, ML is chosen [7]. One of the most crucial handicaps in current ML is the lack of *model traceability*. However, the high predictive power of these methods promotes their use for biological applications. Although, such methods are not exempt from risks.

In the remaining of this section, we discuss the relevance of desirable properties of ML models followed by the current limitations of biomedical ML solutions that lack some of these properties, leading to systematic errors and validity issues. Next, we delve into the details regarding the origin and typology of such systematic errors and the limitations of the performance evaluation of ML models, which has proven unable to express the model's phenomenological fidelity and data validity. These issues are illustrated with examples. Finally, we describe methodological tools to help preventing such systematic issues through the development of the ML models and their future life after deployment.

2.1   The Desirable Properties of ML Models

Like any other model, ML models are purpose dependent. Therefore, there is no exhaustive set of sufficient and necessary conditions to define what a good model is. Nonetheless, there are some common criteria than can be balanced to achieve a good compromise. In 2019, the European Union's (EU) High-Level Expert Group on Artificial Intelligence (AI HLEG) released the "Ethics guidelines for trustworthy AI" prescribing four ethical principles: "respect for human autonomy, prevention of harm, fairness and explicability" accompanied with seven key requirements: (1) human agency and oversight, (2) technical robustness and safety, (3) privacy and data governance, (4) transparency, (5) diversity, non-discrimination, and fairness, (6) environmental and societal well-being, and (7) accountability [8]. To achieve these goals, models must feature desirable properties such as model traceability, robustness and generalisation.

Model Traceability. Producing trustworthy solutions requires traceability, which involves providing a detailed account of data provenance and the design decisions involved in the production of the model. This requirement concerns transparency and explainability, providing rich documentation of the problem assumptions, data collection, labelling, cleaning, model selection, model training, evaluation methods, deployment work and model monitoring. Of course, the particular actions for model traceability depend on the model and its context. An indirect requirement for proper model traceability is data accountability [9] which demands rigorous documentation of the data generation process, its limitations and the underlying assumptions that shape the data. Moreover, traceability intersects with goals such as repeatability, reproducibility and replicability [10]. All in all, the final aim is to successfully reproduce and explain model outcomes and backtrack the prediction process.

Robustness. Biological solutions usually involve multi-modal information, which often involve different sources of data, entailing different kinds of noisy data with peculiar distributions (e.g., non-normality). Such particularities call for robust models able to handle potential outliers and noisy data [11]. ML models are robust when their output remains consistently accurate even after drastic changes in one or more of the input's independent variables (features). Or in other words, the testing error has to be consistent with the training error of the model. For instance, ideally, a model for image classification should be robust to natural distribution shifts in the input, such as changes due to lighting conditions, scene compositions, etc.

Generalisation. Tightly related to the previous virtue, generalisation relates to the model performance in unseen instances drawn from the same distribution as the training set. Note that this does not necessarily refer to noisy data. Although evaluating ML models on non-overlapping test data is a common technique to assess internal validity, it does not reflect the resilience to outliers, noisy data and transferability to other scenarios.

Currently, when we identify a performance change of a ML model on a new dataset, there is few we can do without additional information (e.g., documentation, metadata) that may help us explain why the performance is different. Was the data acquired differently? (e.g., different devices, protocols). Is the context of the sampling units different? (e.g., patients from a region with different disease incidence). Providing the research assumptions encoded as causal diagrams as well as dataset meta-information together with a traceable model may help understanding issues in robustness and generalisation that may arise both during the development of solutions or during the life cycle of a solution already deployed in a real-scenario.

2.2   Current Limitations in Biomedical ML Solutions

The number of scientific works using ML techniques has increased exponentially over the last years, and it is progressively translating into real-world applications, including high-stakes domains such as health, conservation, employment, education or justice. High-stakes AI domains are characterised by their significance and lasting impact on both individuals and society [12].

Unfortunately, despite these models report excellent results during their model training and testing steps, there are notorious cases where the accuracy dropped significantly during their real application, with harmful repercussions when affecting high-stakes domains. Systematic errors have been reported in commercial software employing ML models for tracking, face detection, criminal justice and hiring recommendations. Such errors include systematic biases against concrete populations [13] and limitations in model generalizability and transportability which are well-known issues in biomedical applications [14].

2.3   Origin and Error Types

The reasons and solutions are complex, but one of the most notable factors is the dataset composition, including the consideration and intervention to avoid undesirable biases and potential confounders. Although confounders are not biases *per se*, neglecting control for them may produce biased estimates. Simply put, the main error sources fall into two subtypes of biases. On one side, *conveyance of systematic biases in the datasets*. On the other side, *biases induced during the collection, annotation, preprocessing, and learning strategies*.

Conveyance of Systematic Biases in the Datasets. These patterns represent actual real-world bias that we do not want to convey in the data. For instance, a dataset may reflect an unfair systematic historical discrimination against a particular group of people that may undesirably be perpetuated or even amplified if is not controlled for [13].

Biases Induced During the Collection, Annotation, Preprocessing, and Learning Strategies. In this case, biases arise from one or multiple steps in the data pipeline, e.g., the data collection process might be biased, or the training process may *wrongly employ features*, producing a biased model [15], e.g., in CNN applications [16,17], image features that we may think to have a stronger predictive power for the model may not be as relevant as other hidden features we are unable to see such as noise patterns that hint different acquisition devices, acting as a shortcut to detect two groups of sampling units (e.g., healthy v. control). See Sect. 2.4 for a detailed example.

Although the origin and impact of both type of errors is different, the solution to both involves improving model *traceability* in different ways for a better understanding of the model's decisions. Efforts in this direction are applied in the whole MLOps pipeline, i.e. the automation and monitoring at all steps of ML system development and deployment, including integration, testing, releasing, deployment and infrastructure management [18,19]. This paper focuses on the induced biases as they are an important concern in biological studies. A special focus is put on dataset documentation of both the acquisition process and metadata as a way to reduce the risk of such undesirable biases. Additionally, encoding the assumptions taken during study design is key to spot incompatibilities across datasets, models and studies.

2.4 Limitations of the Current Evaluation System

At this point, one may wonder how models suffering from such issues could satisfy all the requirements needed for their deployment. This issue is explained to a large extent because the metrics currently employed by practitioners (e.g., F1, accuracy, AUC, Matthews correlation coefficient) assess the goodness of the model fitting the data but do not express the phenomenological fidelity of the model and the validity of the data. The phenomenological fidelity refers to the representation of the modelled phenomena. For instance, although diseases can co-exist, many ML solutions designed for disease diagnosis employ multi-class classification approaches which do not properly represent the hypothesis space of the problem to be solved, e.g., diagnosis, since diseases may have more than one etiology, and one etiology can lead to more than one disease [20]. On the other side, the validity of the data indicates how well the data captures the phenomena in order to explain it [12], because apparent performance does not necessarily imply proper phenomena modelling [21].

Current applications measure how good the models perform on the test data, which generally is a subdivision of the same dataset or a dataset collected under similar conditions. Such a score does not express how well the model captures the behaviour of the real phenomena, for which data is just an approximate representation of reality, e.g., they do not measure whether all the event variations are considered or if the capturing methods have enough sensitivity.

Moreover, the 'black box' nature of current ML models hinders transparency regarding the features or combination of features employed during predictions. Equally dangerous is the fact that models are not aware of their surrounding context. Models best hope is to expect their target data resembles that from their training period [20]. Without traceability capabilities and context information of its deployment environment, we cannot properly understand and debug any issues that may arise when a model is deployed in the real-world.

While common ML safe practices like cross-validation or class imbalance control aim to minimise model issues such as model over-fitting, **their use draws**

from the premise that data is a solid representation of the modelled phenomena. Such practices cannot overcome data collection issues, leading to poor consideration of the dataset acquisition process, metadata information and documentation work.

These issues have been proven to impact ML projects, e.g., decreasing the accuracy of IBM's cancer treatment AI solution and causing that Google Flu Trends underestimates a flu peak by 140% [12]. In another example, Maguolo et al. show how the lack of explainability and transparency may hinder the detection of drastic cases of apparent performance with a total lack of phenomena modelling. The authors discuss the validity of testing protocols in most papers dealing with the automatic diagnosis of COVID-19. For that, the authors removed the lungs from X-Ray images and retrained the models, obtaining a similar apparent good performance. The authors conclude that "models might be biased and learn to predict features that depend more on the source of the dataset than they do on the relevant medical information" [21]. Therefore, models may use undesirable bias or exploit uncontrolled confounders during the training. Proper documentation of the acquisition differences may enable rapid comparison of datasets to better spot potential sources of errors during training as well as sudden changes in input data in production settings.

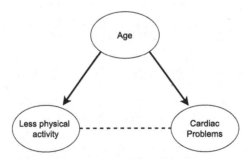

Fig. 1. DAG depicting a scenario with a confounding factor (age) acting (solid lines) in both the predictor and outcome (relationship depicted with a dashed line).

2.5   Helping Methodological Tools

Biological solutions are diverse and there is not a single tool or methodology that may suit all scenarios, studies, settings and research problems. However, every scientist, regardless of the area, shares a common toolkit of methodologies including calibration of instruments, reproducibility, journaling (either lab notebooks or code version control), and so forth. Below we share two methodological practices that may prevent aforementioned issues.

Better Dataset Documentation. The first step to avoid, or detect, potential bias in datasets and models is to improve the documentation of the **dataset generation process**. This is a crucial step since, currently, there is a plethora of techniques to control or correct in case of potential bias or confounders. However most of them assume previous knowledge of such elements [22]. Some of the most common sources of induced bias include unknown confounders as well as selection, acquisition, and annotation biases resulting from **non documented assumptions**. As shown in the directed acyclic graph (DAG) from Fig. 1, confounders (age) are variables that affect both the potential predictor variable (physical activity) and the outcome (cardiac problems). When the presence of confounders is unknown and in lack of experiments specifically designed to minimise them (e.g., randomised controlled trials), we cannot control for them. Uncontrolled confounders lead to conclude that a given feature may be a strong predictor of the outcome when in reality the association is spurious. Moreover, such association may not hold anymore when the sample comes from a different setting where the confounder is differently expressed, e.g. income level may play a different role on diabetes treatment depending on the country [23]. When the model learns spurious associations between predictors and outcomes, an undetected overfitted model is produced, resulting in poor generalisation capabilities that eventually unveil during its translation into real-world settings [14]. Before ML-based analysis expansion, researchers generally employed statistical modelling of biological processes to make inferences from observational data. Generally, in statistical modelling, there is a tight control of potential confounders. This close analysis allows to include functional assumptions that affect the relationships between variables.

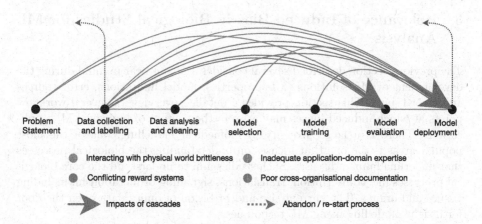

Fig. 2. Diagram depicting data cascades re-drawn from [12]. Thick red arrows show the point where compounding effects become visible. Dotted red arrows depict abandoning or restarting of the ML data process (Color figure online).

Data Cascades. Recently, the concept of data cascades (DC) was presented as one of the main issues in the current life-cycle of an ML system [12]. DCs are defined as compounding events provoking negative downstream impact from data issues causing a technical debt over time. DC describes and identifies how induced biases are generated during the design and data collection process: from the problem statement, dataset collection, data labelling, data analysis and cleaning; as well as model selection, model training, model evaluation until model deployment. This is represented in Fig. 2, which depicts several steps of the development process in which issues may arise, compromising the model. In essence, these cascading issues are similar to those that may arise in the traditional process of drug development, such as translational failures using animal models or issues in patient stratification [24]. If any of these issues (both in drug development or ML development) are not spot at the right stage, e.g., a translation failure between animal models and humans during the pre-clinical stage or a wrong assumption during the problem statement, the cost and potential harm in later stages can be disastrous. Therefore, pro-active work to spot weak points in the workflow can help dividing the research work into milestones at which documentation and quality control are evaluated to prevent dragging such errors to later stages.

Appropriate encoding of research assumptions, metadata information and proper workflow design can help improving replication, reproducibility or assess suitability of models to new data and vice-versa. Bearing all the previous concerns, intervention in the developing and reporting systems of ML-based solutions must be addressed before its translation into real-world settings.

3   Relevance of Induced Bias in Biological Studies for ML Analysis

The previous section described several considerations to bear in mind during the development of ML solutions (ML properties, model limitations, error sources and error types). This section provides a detailed overview of several workflow points at which induced biases may arise in the context of biological ML applications. There is a large diversity in biomedical modelling, one of the most popular areas is the one that focuses on understanding the biological processes that underline human diseases. Such studies aim to understand the core biological processes like transcription, translations, signalling or metabolism, including tissues and organs. The current size and precision of omics data open the door for insight modelling using ML techniques.

There are multiple areas where ML has been successfully employed in biological research, such as gene expression, microRNA binding, protein-protein interaction, single-cell data, metagenomics and sequencing. In consequence, this has impacted many different areas expanding our knowledge in diverse areas such as neurobiology, cancer biology, and immunology. Further research is expected to shake the clinical landscape improving clinical decision-making, predicting

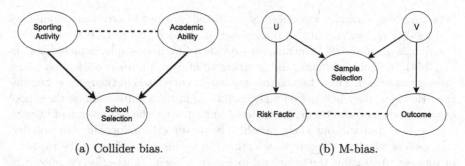

Fig. 3. Directed arrows indicate causal effects and dotted lines indicate induced associations. (a) shows a scenario in which collider bias could distort the estimate of the causal effect of sporting activity on the academic ability. As shown in (b), the relation between the two associated variables can be indirect, with the risk factor and the outcome being indirectly associated with sample selection through unmeasured confounding variables (U and V).

patient trajectories, and boosting clinical trial efficiency and drug development [25].

The **batch effect** (BE) is a known source of confounding in the area of biology. The BE refers to the different factors when comparing sample lots that affect the measurements masking the biological variation impact. The BE is the consequence of different laboratory conditions, reagent lots, machine calibration, software and even personnel differences, e.g., a strong laboratory-specific effect has been reported when comparing multiple micro-array experiments. Another example concerns gene expression studies, in which large variations are associated with the data processing and the specific settings of micro-array work. Consequently, several papers, including relevant studies published in high-impact journals, were retracted [26] on the basis of such errors.

Such issues are usually addressed during the experimental design thanks to randomisation, stratification, replications and inclusion of both positive and negative controls. However, dataset reuse and dataset mix (where datasets are often produced with different settings) may impede controlling for such factors. Therefore, this is a lurking problem in biological data analysis and ML solutions employing mixed high-throughput datasets. Finally, although the BE often relates to the preparation and measure conditions of the samples, other induced bias may arise in the subsequent pipeline steps in the form of data cascades.

Practices leading to **collider bias** represent yet another source of bias. A collider entails a variable that is influenced by two other variables, i.e. collider bias occurs when an exposure and outcome (or factors causing these) each influence a common third variable [27]. The associations induced by collider bias are properties of the sample, rather than the individuals that comprise such a sample. Therefore, such associations fail to generalise beyond the sample and may be inaccurate even within the sample, threatening validity, e.g., when the factors affecting sample selection also affect the variables of interest, the relationships between these variables may become distorted, leading to erroneous inferences and modelling. An account of the impact of collider bias in COVID-19 understanding is given in [28]. The authors provide a simple example, depicted here in Fig. 3 (a), in which academic and sporting abilities can influence selection into a prestigious school. These two factors are barely correlated in the general population. However, they become strongly correlated in the sample because the school enrolment depends on them. As depicted in Fig. 3 (b), the association of interest can be distorted without their variables being directly influencing the collider, e.g., factors affecting the sample selection can themselves influence the variables of interest, distorting the relationship between them. This effect is known as **M-bias**.

Of course, the effect of collider bias is not limited to observational studies and can as well be conveyed to any ML solution trained on biased datasets of this nature.

As depicted in Fig. 4, there are five general steps from which these potentially induced biases can arise and have cascading consequences (see DC in § 2.5). In first place, the biological source may showcase variations in population, disease penetrance, phenotypic manifestation, environmental conditions or sample techniques may induce biases during human sampling. Next, human or machine sample preparation may be sensitive to the machines employed, reactants, protocol settings and in-house calibration. In the same line, different settings conditions may affect the signal measurement. Then data analysis is conditioned by the approach employed for data cleaning, normalisation and labelling. Finally, data sharing is not exempt from issues if decisions are taken to modify or remove features or statistical units before the data distribution.

All of these issues may result in technical and research debt in later stages of the scientific programmes.

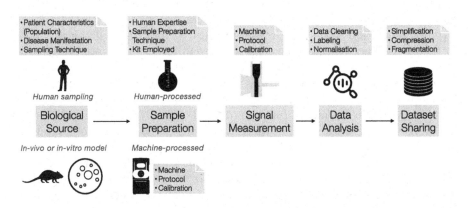

Fig. 4. General workflow diagram showcasing the generation process of biological datasets. Gray boxes contain potential sources of induced bias

4   An Approach to Overcome the Limitations: Accompanying Metadata with Causal Information

Open science and open innovation allow fulfilling a basic principle of science, reproducibility. The main principles include open code, open data and open publications. But while open data allows reproducing the reported results, ensuring data reusability entails proper description of the whole generation process. At this point, it is clear that datasets must be paired with proper documentation and accompanied with additional information to automatise dataset and pipeline meta-analysis. These are indeed among the principles of relevant guidelines such as DOME [6]. However, encoding such information together with the dataset is not always possible, often requiring sidecar files better suited to express dataset properties differently. Metadata has emerged as a crucial component for reproducibility in the research life cycle [29]. Additionally, the full potential of metadata is still open to unexplored opportunities associated with the area of biological ML [6]. FAIR principles reflect the need of reusability and interoperability, suggesting extensive documentation to satisfy data management and stewardship needs. Similarly, DOME guidelines aim for proper data provenance and safe model evaluation. However, such principles do not demand further metadata encoding of the causal assumptions made during the data collection process and the intentions of the original study for which it was collected.

The final aim of such metadata should be to convey the data generating process enabling its comparison across datasets to identify differences in the generation process and inform of potential induced biases as the first step for its control. To ensure its correct comparison the metadata should be standardised. The metadata standardisation is already present in domain-specific repositories such as Genbank or UniProt which are highly curated and include specific metadata. Domain-specific metadata standards include DICOM (Digital Imaging and Communications in Medicine), FHIR (Fast Healthcare Interoperability Resources), Functional Annotation of ANimal Genomes (FAANG) and Observational Health Data Sciences and Informatics (OHDSI). However, general-purpose solutions are still scarce [5] with ISA Commons [30] as the prominent model to describe metadata relating to the provenance of samples and data.

However, none of these metadata formats and models enable the encoding of causal graphs representing the assumptions considered during the data collection process or study design. Such additional information could prevent issues in which the modeller is unaware of known confounders (which might be taken into account during the data acquisition but not properly documented) between the variables, with unexpected consequences.

Additionally, an inappropriate split may break the assumption that data is independent and identically distributed, in other words, that all samples stem from the same generative process which has no memory of past generated samples. For instance, a medical dataset containing multiple samples from the same patients without stating the patient id (or another patient dependent variable), precluding group-wise division of the dataset. In this example, the training process could be compromised due to potential data leakage caused by the presence of samples from the same patient in both the train and test sets. Similar to group-wise split issues, if the samples stem from a time-dependent process, a time-wise scheme is at hand, otherwise data leakage may also occur if instances of different moments in time are present in both train and test sets. In any case, such data generative process must be properly documented beforehand.

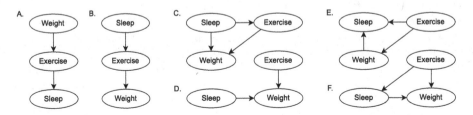

Fig. 5. Sample causal diagrams showcasing some combinations for 3 variables. In C, sleep acts as a confounder of both exercise and weight, while in E its role is a collider. Finally, in F, sleep acts as a mediator.

4.1 Incorporating Causal Information

As previously discussed, causal diagrams enable encoding of assumptions over the modelled phenomena. Causal diagrams summarise existing knowledge in the form of nodes and edges which convey causal relationships between variables (nodes). First, such knowledge is simplified into assumptions which are then used to construct causal diagrams, or causal models. Again, such assumptions are encoded as missing edges and the direction of present edges, e.g., exercise reduces weight. Causal diagrams enable calculating *estimands* prior to data collection. Given the causal query P(W|do(E)) (read as the causal effect of exercise in weight), and thanks to the *estimand* P(W|S,E)·P(S) from diagram C from Fig. 5, we know that the question cannot be answered without sleep data. This causal information could be used to select only those datasets that have sleep data. Then, metadata could provide information regarding sleep data acquisition to ensure compatibility with previous data or across datasets when merging datasets. Note that for diagram E, weight is assumed not to depend on sleep, hence no sleep data would be required in such scenario.

Data Alone is Not Enough. Importantly, such causal information cannot be extracted from tabular or image data, however, we need to convey (when our knowledge allows it) how variables relate. For instance, we can track the length of a building shadow B as a function of the sun's elevation S and building height h, but tabular data alone will not be able to express the direction of the causal relationship between S and B. However, since we know that the former affects the latter and not vice-versa, we can encode such knowledge in a causal model. This is even more important when knowledge is unclear but assumptions are nevertheless taken, e.g., does obesity increase asthma or vice-versa? are there mediators? In such cases, it is essential to convey the research assumptions taken during the construction of a solution. These situations can also arise in biological system modelling for the understanding of disease molecular mechanisms, e.g., an elevation of a determined protein and a concrete gene may be observed under two different conditions. This association may be bi-directional, i.e., the protein may act as a regulator of the gene expression, or the gene expression may stimulate protein generation.

Acquisition metadata can also help raising warnings, e.g., if acquisition devices are replaced, which may require certain model re-calibration to prevent errors. Conversely, if the assumptions change as a consequence of new knowledge introduction, warnings may be raised regarding the data, or new incompatibilities may be found, e.g., if we learn sleep to affect both exercise and weight, the system may check whether sleep data acquisition methods are compatible thanks to the metadata information.

Making Causal Information Actionable. Causal diagrams can be encoded in file formats such as the graph description language DOT or Graph Modelling Language (GML). Python libraries such as Microsoft's DoWhy [31] and IBM's causallib [32] allow working with such causal models enabling calculation of *estimands* and estimate of causal effects. Both libraries provide detailed examples in the form of Jupyter Notebooks to help getting familiar with their tools. Such tools can be employed to work with causal models of high level knowledge in a similar way Systems Biology Markup Language (SBML) allows to represent low-level biological phenomena such as cell signalling pathways [33,34]. In a similar way to SBML models, causal models could be labelled with standard identifiers such as those available at https://identifiers.org/, which includes more than 700 different namespaces [35] such as the international convention of diseases (ICD) or SNOMED collection of medical terminology.

Together with causal information, and according to the above, metadata (following either general-purpose or domain-specific standards) should at least comply with the following principles to enable actionable metadata application. **Interoperability**, through automatic metadata generation during data collection that eases machine-machine interaction. **Usability**, enabling easy-to-use integration when human input is required. **Adherence**, supplying an interface to which general and domain-specific standards may adhere. **Integrative**, employing already existing guidelines. **Privacy**, providing features to comply with current data protection frameworks (such as the EU General Data Protection Regulation or the California Consumer Privacy Act).

5   Conclusion

In this paper we present the general issues affecting ML solutions for high-stakes domains, such as biomedical research, caused by induced biases derived from the collection, annotation and preprocessing stages of the ML model production pipeline.

In particular, we address the lack of information context which enables ML models to learn spurious associations between variables that might be affected by endogenous confounding factors derived from the particularities of the pipeline setting (such as instrument noise, laboratory protocols).

This issue is amplified in high-throughput scenarios when comparing sample lots generated in different settings. Our proposed approach involves bringing data work to the foreground of the ML model production pipeline. Increased domain knowledge, data excellence incentives and improved feedback channels in the AI data life-cycle are good starting points [12]. However, such goals must be translated into material actions.

In this paper we propose increased documentation of the dataset generation process as an essential safety practice. This includes the use of dataset-wise standardised metadata and incorporation of causal relationship information regarding dataset variables.

In this sense, we note the scarcity of general-purpose metadata standards but emphasise the availability of purpose-dependent metadata standards such as DICOM. Rather than proposing a standard for metadata, we recommend to exploit existing metadata standards and accompany such information with causal assumptions, integrating these in the workflow to both strengthen the quality of model development and improve audit and monitor deployed solutions.

We believe the inclusion of metadata is well suited for the particularities of biological datasets and may ease dataset mixture and pipeline comparison. The practice of including causal information in the documentation could prevent confounding effects by encoding the assumptions concerning the dataset generation process. Together, this extra information can be used to audit dataset compatibility and model suitability.

Thus, existing standards are yet to be fully exploited for the production of standardised documentation and metadata, which may not only ease the data work but also open the door for actionable-metadata and incorporation of causal relationships during the model training. We believe these strategies will help mitigating potential risks of ML solutions in real-world scenarios both during development and deployment stages.

Acknowledgments. The authors would like to thank Andreas Husch and Matias Bossa for their support and review efforts. Beatriz Garcia Santa Cruz work is supported by the FNR-PRIDE17/12244779/PARK-QC and Pelican award from the Fondation du Pelican de Mie et Pierre Hippert-Faber.

References

1. A survey on deep learning in medical image analysis. Med. Image Anal. **42**, 60–88 (2017)
2. Angermueller, C., Pärnamaa, T., Parts, L., Stegle, O.: Deep learning for computational biology. Mol. Syst. Biol. **12**, 878 (2016)

3. Repecka, D., et al.: Expanding functional protein sequence spaces using generative adversarial networks. Nat. Mach. Intell. **3**, 324–333 (2021)
4. Xu, C., Jackson, S.: Machine learning and complex biological data. Genome Biol. **20**, 1–4 (2019)
5. Wilkinson, M., et al.: The FAIR guiding principles for scientific data management and stewardship. Sci. Data **3**, 1–9 (2016)
6. Walsh, I., et al.: DOME: recommendations for supervised machine learning validation in biology. Nat. Methods **18**, 1122–1127 (2021)
7. Bzdok, D., Altman, N., Krzywinski, M.: Statistics versus machine learning. Natu. Methods **15**, 233 (2018)
8. Smuha, N.: The EU approach to ethics guidelines for trustworthy artificial intelligence. Comput. Law Rev. Int. **20**, 97–106 (2019)
9. Hutchinson, B., et al.: Towards accountability for machine learning datasets: Practices from software engineering and infrastructure. In: Proceedings of the 2021 ACM Conference on Fairness, Accountability, and Transparency, pp. 560–575 (2021)
10. Mora-Cantallops, M., Sanchez-Alonso, S., Garcia-Barriocanal, E., Sicilia, M.: Traceability for trustworthy AI: a review of models and tools. Big Data Cogn. Comput. **5**, 20 (2021)
11. Paschali, M., Conjeti, S., Navarro, F., Navab, N.: Generalizability vs. robustness: adversarial examples for medical imaging. ArXiv Preprint ArXiv:1804.00504 (2018)
12. Sambasivan, N., Kapania, S., Highfill, H., Akrong, D., Paritosh, P., Aroyo, L.: Everyone wants to do the model work, not the data work: data cascades in high-stakes AI. In: Proceedings of the 2021 CHI Conference on Human Factors in Computing Systems, pp. 1–15 (2021)
13. Mitchell, M., et al.: Model cards for model reporting. In: Proceedings Of The Conference On Fairness, Accountability, And Transparency, pp. 220–229 (2019)
14. Santa Cruz, B., Bossa, M., Sölter, J., Husch, A.: Public Covid-19 X-ray datasets and their impact on model bias-a systematic review of a significant problem. Med. Image Anal. **74**, 102225 (2021)
15. Castro, D., Walker, I., Glocker, B.: Causality matters in medical imaging. Nat. Commun. **11**, 1–10 (2020)
16. Zhu, Y., et al.: Converting tabular data into images for deep learning with convolutional neural networks. Sci. Rep. **11**, 1–11 (2021)
17. Bazgir, O., Zhang, R., Dhruba, S., Rahman, R., Ghosh, S., Pal, R.: Representation of features as images with neighborhood dependencies for compatibility with convolutional neural networks. Nat. Commun. **11**, 1–13 (2020)
18. Mäkinen, S., Skogström, H., Laaksonen, E., Mikkonen, T.: Who needs MLOps: what data scientists seek to accomplish and how can MLOps Help?. In: 2021 IEEE/ACM 1st Workshop on AI Engineering-Software Engineering For AI (WAIN), pp. 109–112 (2021)
19. Sweenor, D., Hillion, S., Rope, D., Kannabiran, D., Hill, T., O'Connell, M.: ML Ops: Operationalizing Data Science. O'Reilly Media, Incorporated (2020)
20. Vega, C.: From Hume to Wuhan: an epistemological journey on the problem of induction in COVID-19 machine learning models and its impact upon medical research. IEEE Access. **9**, 97243–97250 (2021)
21. Maguolo, G., Nanni, L.: A critic evaluation of methods for COVID-19 automatic detection from X-ray images. Inf. Fusion **76**, 1–7 (2021)
22. VanderWeele, T.: Principles of confounder selection. Eur. J. Epidemiol. **34**, 211–219 (2019)

23. Beran, D., Lazo-Porras, M., Mba, C., Mbanya, J.: A global perspective on the issue of access to insulin. Diabetologia **64**, 954–962 (2021)
24. Altevogt, B., Davis, M., Pankevich, D., Norris, S.: Improving and Accelerating Therapeutic Development for Nervous System Disorders: Workshop Summary. National Academies Press, Washington (2014)
25. Ching, T., et al.: Opportunities and obstacles for deep learning in biology and medicine. J. Roy. Soc. Interface. **15**, 20170387 (2018)
26. Leek, J., et al.: Tackling the widespread and critical impact of batch effects in high-throughput data. Nat. Rev. Genet. **11**, 733–739 (2010)
27. Holmberg, M., Andersen, L.: Collider Bias. JAMA **327**, 1282–1283 (2022)
28. Griffith, G., et al.: Others collider bias undermines our understanding of COVID-19 disease risk and severity. Nat. Commun. **11**, 1–12 (2020)
29. Leipzig, J., Nüst, D., Hoyt, C., Ram, K., Greenberg, J.: The role of metadata in reproducible computational research. Patterns. **2**, 100322 (2021)
30. Sansone, S., et al.: Toward interoperable bioscience data. Nat. Genet. **44**, 121–126 (2012)
31. Sharma, A., Kiciman, E.: DoWhy: an end-to-end library for causal inference. ArXiv Preprint ArXiv:2011.04216 (2020)
32. Shimoni, Y., et al.: An evaluation toolkit to guide model selection and cohort definition in causal inference. ArXiv Preprint ArXiv:1906.00442 (2019)
33. Keating, S., et al.: SBML Level 3: an extensible format for the exchange and reuse of biological models. Mol. Syst. Biol. **16**, e9110 (2020)
34. Touré, V., Flobak, A., Niarakis, A., Vercruysse, S., Kuiper, M.: The status of causality in biological databases: data resources and data retrieval possibilities to support logical modeling. Briefings Bioinform. **22**, bbaa390 (2021)
35. Juty, N., Le Novere, N., Laibe, C.: Identifiers. org and MIRIAM Registry: community resources to provide persistent identification. Nucleic Acids Res. **40**, D580–D586 (2012)

Deep Recurrent Neural Networks for the Generation of Synthetic Coronavirus Spike Protein Sequences

Lisa C. Crossman[1,2](✉) [iD]

[1] SequenceAnalysis.co.uk, NRP Innovation Centre, Norwich Research Park, Norwich, UK
l.crossman@uea.ac.uk
[2] School of Biological Sciences, University of East Anglia, Norwich, Norfolk, UK

Abstract. With the advent of deep learning techniques for text generation, comes the possibility of generating fully simulated or synthetic genomes. For this study, the dataset of interest is that of coronaviruses. Coronaviridae are a family of positive-sense RNA viruses capable of infecting humans and animals. These viruses usually cause mild to moderate upper respiratory tract infection; however, they can also cause more severe symptoms, gastrointestinal and central nervous system diseases. The viruses are capable of flexibly adapting to new environments, hence health threats from coronavirus are constant and long-term. Immunogenic spike proteins are glycoproteins found on the surface of Coronaviridae particles that mediate entry to host cells. The aim of this study was to train deep learning neural networks to produce simulated spike protein sequences, which may be able to aid in knowledge and/or vaccine design by creating alternative possible spike sequences that could arise from zoonotic sources in future. Deep learning recurrent neural networks (RNN) were trained to provide computer-simulated coronavirus spike protein sequences in the style of previously known sequences and examine their characteristics. The deep generative model was created as a recurrent neural network employing text embedding and gated recurrent unit layers in TensorFlow Keras. Training used a dataset of alpha, beta, gamma, and delta coronavirus spike sequences. In a set of 100 simulated sequences, all 100 had most significant BLAST matches to Spike proteins in searches against NCBI non-redundant dataset (NR) and possessed the expected Pfam domain matches. Simulated sequences from the neural network may be able to guide us with future prospective targets for vaccine discovery in advance of a potential novel zoonosis.

Keywords: Coronavirus · Deep learning · Neural networks

1 Introduction

1.1 Coronaviridae

Coronaviridae are a family of large, enveloped single-stranded positive-sense RNA viruses encompassing alpha, beta, gamma, and delta coronavirus divisions as well as unclassified divisions in the sequence databases. The genome is packed inside a helical

© The Author(s), under exclusive license to Springer Nature Switzerland AG 2022
D. Chicco et al. (Eds.): CIBB 2021, LNBI 13483, pp. 217–226, 2022.
https://doi.org/10.1007/978-3-031-20837-9_17

capsid and is further surrounded by an envelope. The spike protein forms large pro-trusions from the virus surface, giving the coronaviruses the appearance of wearing a 'crown' under electron microscopy. Coronaviruses can infect a wide range of different animals and usually cause mild to moderate upper-respiratory tract illnesses, however they can also cause severe respiratory infections as well as gastrointestinal and central nervous system diseases. Coronaviruses circulate among humans and animals such as bats, pigs, camels, and cats. Recent zoonoses include severe acute respiratory syndrome Coronavirus (SARS-CoV), which emerged in November 2002 and became effectively extinct by 2004 [1]. Another zoonosis, Middle East Respiratory Syndrome (MERS-CoV) was believed to be transmitted from an animal reservoir in camels in 2012 [2]. In veterinary terms, economically important CoV exist such as porcine epidemic diarrhoea coronavirus (PEDV) which lead to an extremely high fatality rate in piglets [3]. The coronavirus SARS-CoV-2 emerged from China in 2019 [4] and was declared a pan-demic during the first quarter of 2020 with an extremely high requirement for a vaccine to be provided in a short timeframe. The Spike protein is a multifunctional viral protein found on the outside of the SARS-CoV-2 virus particle (Fig. 1).

Fig. 1. Spike protein structure from SARS-CoV-2. The S1 fragment is shown in magenta, the S2 fragment is shown in red, with glycosylation as lighter hues. The receptor binding domain (RBD) is located at the top of the molecule, whilst the S1 and S2 fragments form part of a complex with a membrane-spanning segment. Image adapted from D. Goodsell and RCSB PDB [5]. (Color figure online)

Spike protein initially binds a host cell receptor though its S1 subunit and fuses viral and host membranes through its S2 subunit. In addition to mediating entry, the spike is a critical determinant of viral host range and a major inducer of host immune responses [6]. Due to the key role of the Spike (S) protein, it is the main target for antibody-mediated neutralization [7].

1.2 Recurrent Neural Networks

Deep learning is a subset of artificial intelligence employing neural networks. The recur-rent neural network (RNN) is a type of neural network usually used for text encoding implementations, mainly through whole word encoding and the bag of words concept.

The recurrent neural network (RNN) is trained on a set of sequences using an optimization algorithm with estimations of gradient descent combined with backpropagation through time. The RNN has the potential to consider previously seen data such as the character or word that came before the current time step using units such as long short-term memory cells (LSTM) or gated recurrent units (GRU). The GRU is a variant of LSTM with a forget gate but having fewer parameters than LSTM as it lacks an output gate [8]. GRU performance is similar to LSTM but can be enhanced on some datasets.

In 2007, Hochreiter, Heusel and Obermayer proposed the use of LSTM for protein homology detection [9], commenting that LSTM is capable of automatically extracting local and global sequence statistics like hydrophobicity, polarity, volume and polarizability and combining them with a pattern. The results included extraction of feature dependencies that were not detected with common bioinformatic techniques. In this study, we investigate whether GRU is capable of learning these features in the context of generating synthetic sequences.

2   Methods

2.1   Recurrent Neural Network (RNN) Architecture

In creating the model described in this study, character encoding was used on the sequences in the training set. Alternative model architectures were considered which included either two layers of GRU, a single bidirectional layer of GRU, or two bidirectional layers of GRU. In addition, either a single dense layer was used as output, or two dense layers, with the first dense layer having half the number of RNN units (512). Model architecture changes also included swapping GRU for LSTM. However, the model showing the best results as judged by bioinformatic analysis of the output synthetic sequences was composed of a single embedding layer and a gated recurrent unit (GRU) with 1024 RNN units followed by a dense linear layer (Fig. 2).

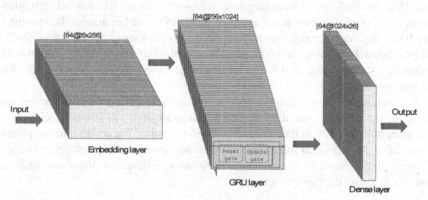

Fig. 2. Simple recurrent neural network trained for the production of synthetic protein sequences of Spike protein. The vocabulary size was 26, including each of the protein single letter codes, plus newline and space characters, the embedding dimension was 256, with a RNN units parameter of 1024. During prediction, after loading the trained model, the model is reset to a batch size of one and prediction is carried out one character at a time starting with the seed text.

The model was trained in Tensorflow 2.1.0 with Keras using an Adam optimizer with AMSgrad option and an adaptive learning rate over 15 epochs, where losses fell gradually from an initial 3.259 to 0.266. Learning was terminated after the losses had fallen between 0.2–0.3.

2.2 Coronavirus Training Set

A training dataset was formulated from a wide variety of coronavirus spike protein sequences from alpha, beta, gamma and delta coronaviruses and constituted isolates from many different animals. The total number of spike protein sequences in the training dataset was 2406, encompassing 511 sequences from Human CoV including examples of SARS-CoV-1 and MERS as well as SARS-CoV-2 (hCoV-19), 232 Bovine, 194 Noctilionine (Bat), 106 Porcine and several samples from other animals including camel, Chinese ferret-badger, hedgehog, dog, deer, avian and whale. Downloaded sequences were searched and cleaned to remove poorer quality and partial sequences and subunits resulting in a total of 2295 sequences. All the cleaned data was used in training and the model was evaluated by bioinformatic methods. BLASTP percentage match identities within the dataset had a mean of 79.7, median of 92.5 and an interquartile range of 37.2.

3 Results

3.1 Characteristics of DL Simulated Spike Proteins

To create predictions, the RNN is initially given a short seed protein sequence. The seed sequence can be passed as a random choice from previously sequenced spike proteins or formulated of random choices of amino acids starting with Methionine chosen by the python random library. In this study, given a seed text, the RNN was then able to provide sequences up to the full length of spike protein, a maximum length in the input dataset of 1582 amino acids with a mean length of 1324.4 amino acids. The maximum sequence identity that a simulated sequence achieved in BLAST matches against the training set was 100% sequence identity over 875 amino acids with a temperature scaling value of 1.0 (see below for details) or 100% identity over the full length of the protein with a temperature scaling value of 0.5. The lengths of all the synthesized proteins were fixed at 1588 amino acids.

For preliminary investigations, 100 DL synthesised spike protein sequences were collected. The RNN was initially provided with seed sequences of 16 amino acids chosen at random from the starts of the full dataset of spike proteins. The amino acid complement of the real and synthesized spike proteins in the datasets is as compared below in Fig. 3. Although the amino acid complements show some differences, there are significant similarities across the two datasets.

Sequence Matching
All 100 of the simulated sequences had a significant BLASTP match to Spike protein from one or more coronavirus sequences with BLAST searches of the query sequences against the entire non-redundant database (Fig. 4).

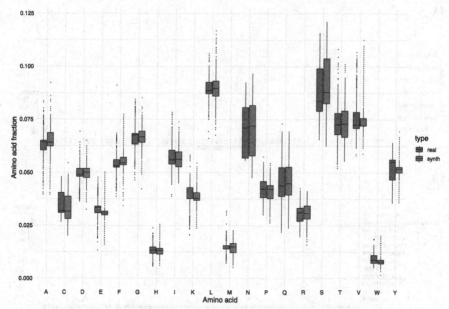

Fig. 3. Comparison of the amino acid composition of the real and simulated proteins.
Boxplot graph showing the amino acid composition of each amino acid as a fraction of the protein sequence in both the real dataset (red) and the synthesized dataset (blue). The amino acid single letter code is shown on the X axis with the fraction of the amino acid in each sequence on the y axis as calculated by Biopython ProtParam module. The 'Real' training dataset comprised 2295 sequences in total with the 'Synth' simulated example dataset containing a matched number of samples. The difference between the Real and Synth amino acid composition datasets was not significant (Mann-Whitney test in R, p-value = 0.35).

```
Query 1  MFVFLVLLPLVSSQCVNLTTRTQLPPAYTNSFTRGVYYPDKVFRSSVLHSTQDLFLPFFS  60
         MFVFLVLLPLVSSQCVNLTTRTQLPPAYTNS TRGVYYPDKVFRSSVLH TQDLFLPFFS
Sbjct 1  MFVFLVLLPLVSSQCVNLTTRTQLPPAYTNSSTRGVYYPDKVFRSSVLHLTQDLFLPFFS  60
```

Fig. 4. Partial BLAST alignment of Spike protein from a simulated query protein against Bat coronavirus RaTG13 [10]

The real spike protein training dataset was clustered and deduplicated resulting in 154 clusters of non-redundant sequences. The 100 simulated sequences were searched with BLASTP against the representative cluster sequences. Figure 5A shows that the best BLAST hits for the first set of simulated sequences covered several distinct clusters. There were several hits to MERS clusters, possibly due to a high representation of MERS sequences in the training set. A second set of 100 simulated sequences were generated that each had an identical seed text of 64 amino acids from the start of SARS-CoV-2 spike. Figure 5B shows that the equivalent BLAST hits on these sequences had a higher number of SARS-CoV matches, as well as Bat SARS-like sequences, although some samples still shared high identities with MERS sequences.

5A

5B

Fig. 5. BLAST matches of simulated sequences against known Spike proteins

Figure 5A shows BLAST searches of the original 100 DL synthesized sequences filtered for matches by length over 200bp and identity over 90%. 5B shows the second set of DL synthesized sequences which were all given identical seed text feeder sequence of 64 amino acids from the start of SARS-CoV-2. This graph is filtered for matches by length over 500 bp and identity over 90%. Highest length matches are represented by the largest diameter circle and darkest colour. However, large circles are generally of interest since the identity cut-off is high. Unfiltered data can be found at the GitHub site as described in the Data Availability section.

Pfam Domain Complements of Simulated Protein Sequences

Significant Pfam domain hits were uncovered on searching the query sequences with HMMER3 against the Pfam-A database. Searches of the synthesized proteins against Pfam_A.hmm database revealed Pfam domains that were expected within a coronavirus spike protein (below).

Table 1 Common Pfam domains and their counts identified within the original 100 simulated sequences which are also found in real Spike proteins, showing that all 100 sequences had C-terminal Spike domains. Other domains were identified in full Pfam-A however, the most common were Corona_S2, Spike_rec_bind and Spike_NTD. Database Pfam-A.SARS-CoV-2 refers to the April 2, 2020 update for SARS-CoV Pfam domains (Xfam Blog https://xfam.wordpress.com/2020/04/02/pfam-sars-cov-2-special-update/).

The resulting Pfam domains compared favourably with the most commonly found domains within the real training dataset of 2504 proteins which had 1781 domain counts of Spike_rec_bind, 2413 Corona_S2, 1052 Spike_NTD and 502 Corona_S1 (Coronavirus S1 glycoprotein domain) among others. According to Pfam Architectures, domain Corona_S2 is found in real spike proteins in the databases together with

Table 1. Pfam Domain Complements in the 100 simulated sequences.

Pfam domain	Pfam database	Full name	Count
Corona_S2	Pfam-A	Coronavirus S2 glycoprotein	100
Corona_S1	Pfam-A	Coronavirus S1 glycoprotein	13
Spike_rec_bind	Pfam-A	Spike receptor binding domain	81
Spike_NTD	Pfam-A	Spike glycoprotein N-terminal	53
CoV_NSP2_C	Pfam-A.SARS-CoV-2	Coronavirus replicase NSP2, C-terminus	6
CoV_S1_C	Pfam-A.SARS-CoV-2	Coronavirus Spike S1, C-terminus	75
bCoV_S1_RBD	Pfam-A.SARS-CoV-2	Betacoronavirus Spike S1, receptor-binding	82
bCoV_S1_N	Pfam-A.SARS-CoV-2	Betacoronavirus-like spike S1, N-terminus	88
CoV_S2	Pfam-A.SARS-CoV-2	Coronavirus Spike glycoprotein S2	100

either Corona_S1, Spike_NTD and Spike_rec_bind, or with just Spike_rec_bind, or with Spike_NTD and 2 x Spike_rec_bind or in some sequences as a standalone domain.

Prediction

During prediction, probabilities are generated for the next character in the sequence of the amino acid single letter alphabet. A parameter, known as temperature, can be used to scale the probabilities of the output distribution. If the temperature value is low, the model will be more confident on predictions which may produce more repetitive text. At a temperature of 0.5, the model was able to reach 100% identity over the full length of Spike SARS glycoprotein with the only differences being in the seed text. The purpose of this study is to provide sequences that are not identical to known sequences so we may find better use of a higher temperature value to provide more diverse text.

A second dataset sample of 100 synthesized sequences was formed by specifically using a seed text of 64 amino acids from the SARS-CoV-2 spike protein for each simulated sequence. When simulated sequences were clustered at the default 90% level of sequence identity, the result was 51 separate clusters in which Cluster 0 had 27 members ranging from 92%–100% identity which corresponded to SARS-CoV-1 type, Cluster 1 had 13 members of 97–100% identity which corresponded to Bat RaTG13/SARS-CoV-2 type, Cluster 38 had 3 members corresponding to MERS type, Cluster 2 had 3 members, Clusters 4, 8 and 19 each had two members, whilst each example of the rest of the dataset clustered separately. Hence, the seed text provided SARS-like hits in several but not all cases. Some sequences were definitively of interest to this study, such as a synthesized protein with 97% full length identity to a Bat beta-coronavirus sequence isolated from *Chaerephon plicata* in Yunnan in 2011 [11]. Further sequences of interest included those with high identity over stretches of the protein sequence to SARS-CoV-1 or SARS-CoV-2, particularly those including hybrid regions. Once the initial predicted protein is finished, the prediction commences a new protein again immediately if the maximum number of characters has not been reached. In some cases there were hybrid matches to parts of sequence from spike proteins in the dataset. A larger dataset of 1000 simulated sequences was generated with the same SARS-CoV seed text as previously.

These simulated sequences were clustered and clusters corresponding to SARS-CoV-1 and SARS-CoV-2 were aligned together with examples of the real spike proteins from SARS-CoV-1 and SARS-CoV-2. Multiple sequence alignments of the binding region indicate that residues important in human ACE2 recognition [12, 13] are broadly conserved across the simulated sequences.

6A

6B

Fig. 6. Sequence Logos from multiple sequence alignments of the spike receptor binding domains (RBD) generated on Weblogo3 [14]. Each logo consists of stacks of symbols, one for each position in the sequence. Colours denote amino acid chemical properties. The height of the stack indicates the relative frequency of each amino acid whilst the width of the stack shows the fraction of symbols in the column (narrow = many gaps). Figure 6A shows the receptor binding domains of all the proteins in the real dataset that possessed a RBD matching that of the SARS-CoV and SARS-CoV-2 clusters in the dataset. Figure 6B shows the RBD of the synthetic dataset alignments that matched those clusters. The starred amino acids are contacting residues with human ACE2 receptor in the RBD of both SARS-CoV and SARS-CoV-2 [13].

4   Conclusions

This study used a comprehensive training set formulated from Coronavirus Spike protein sequences in DL neural networks to produce novel sequence from a short feeder seed text. Novel sequences shared features that can be searched with bioinformatics tools to provide highly significant BLAST and Pfam domain matches. That each of the sequences examined shared matches to Spike protein is exciting and warrants further consideration. Interestingly, in one example, the prediction query was able to correct an unknown amino acid (X) to a G that exactly matched other sequences of that type. It is trivial to generate high numbers of these synthesized sequences from the model, although in some cases the resultant sequence may not represent viable protein. In addition, the training set is only as comprehensive as the initial database, animal CoV sequences may exist elsewhere that are not represented here, and further unknown biases may exist.

Synthesised sequences may find a use in cases of data privacy such as generating synthetic patient data for studies as provided by Synthea (Standard Health Record Collaborative), or to generate further examples of poorly represented data for better statistical analysis. Further applications may include the generation of gene clusters of a particular type and evolution studies. Whilst the CoV model occasionally produced data with large-scale rearrangements, the results indicated that relatively simple neural networks can provide useful synthetic sequences with low compute requirements when trained on a curated database. The potential production of novel sequence by DL is exciting as a future strategy and warrants further consideration.

Data Availability.   The model and source code are available at: https://github.com/LCrossman

References

1. Organization WH: Consensus document on the epidemiology of severe acute respiratory syndrome (SARS). WHO/CDS/CSR/GAR/2003.11 (2003)
2. Zaki, A.M., Van Boheemen, S., Bestebroer, T.M., et al.: Isolation of a novel coronavirus from a man with pneumonia in Saudi Arabia. N. Engl. J. Med. (2012). https://doi.org/10.1056/NEJMoa1211721
3. Zhou, P., Fan, H., Lan, T., et al.: Fatal swine acute diarrhoea syndrome caused by an HKU2-related coronavirus of bat origin. Nature (2018). https://doi.org/10.1038/s41586-018-0010-9
4. Zhu, N., Zhang, D., Wang, W., et al.: A novel coronavirus from patients with pneumonia in China, 2019. N. Engl. J. Med. (2020). https://doi.org/10.1056/NEJMoa2001017
5. Goodsell, D.: Molecule of the Month SARS-CoV-2 Spike (2020). https://doi.org/10.2210/rcsb_pdb/mom_2020_6. http://pdb101.rcsb.org/motm/246. Accessed 14 June 2022
6. Li, F.: Structure, function, and evolution of coronavirus spike proteins. Annu. Rev. Virol. (2016). https://doi.org/10.1146/annurev-virology-110615-042301
7. Zhou, G., Zhao, Q.: Perspectives on therapeutic neutralizing antibodies against the Novel Coronavirus SARS-CoV-2. Int. J. Biol. Sci. (2020). https://doi.org/10.7150/ijbs.45123
8. Cho, K., Van Merriënboer, B., Gulcehre, C., et al.: Learning phrase representations using RNN encoder-decoder for statistical machine translation. In: EMNLP 2014 - 2014 Conference on Empirical Methods in Natural Language Processing, Proceedings of the Conference (2014)
9. Hochreiter, S., Schmidhuber, J.: Long short-term memory. Neural Comput. (1997). https://doi.org/10.1162/neco.1997.9.8.1735

10. Zhou, P., Lou, Y.X., Wang, X.G., et al.: A pneumonia outbreak associated with a new coronavirus of probable bat origin. Nature (2020). https://doi.org/10.1038/s41586-020-2012-7

11. Wu, Z., Yang, L., Ren, X., et al.: ORF8-related genetic evidence for Chinese horseshoe bats as the source of human severe acute respiratory syndrome coronavirus. J. Infect. Dis. (2016). https://doi.org/10.1093/infdis/jiv476

12. Luan, J., Lu, Y., Jin, X., Zhang, L.: Spike protein recognition of mammalian ACE2 predicts the host range and an optimized ACE2 for SARS-CoV-2 infection. Biochem. Biophys. Res. Commun. (2020). https://doi.org/10.1016/j.bbrc.2020.03.047

13. Lan, J., Ge, J., Yu, J., et al.: Structure of the SARS-CoV-2 spike receptor-binding domain bound to the ACE2 receptor. Nature (2020). https://doi.org/10.1038/s41586-020-2180-5

14. Crooks, G.E., Hon, G., Chandonia, J.M., Brenner, S.E.: WebLogo: a sequence logo generator. Genome Res. (2004). https://doi.org/10.1101/gr.849004

Recent Dimensionality Reduction Techniques for High-Dimensional COVID-19 Data

Ioannis L. Dallas[1(✉)], Aristidis G. Vrahatis[2], Sotiris K. Tasoulis[1],
and Vassilis P. Plagianakos[1]

[1] Department of Computer Science and Biomedical Informatics,
University of Thessaly, Lamia, Greece
{idallas,stasoulis,vpp}@uth.gr
[2] Department of Informatics, Ionian University, Corfu, Greece
aris.vrahatis@ionio.gr

Abstract. We are going through the last years of the COVID-19 pandemic, where almost the entire research community has focused on the challenges that constantly arise. From the computational and mathematical perspective, we have to deal with a dataset with ultra-high volume and ultra-high dimensionality in several experimental studies. An indicative example is DNA sequencing technologies, which offer a more realistic picture of human diseases at the molecular biology level. However, these technologies produce data with high complexity and ultra-high dimensionality. On the other hand, dimensionality reduction techniques are the first choice to address this complexity, revealing the hidden data structure in the original multidimensional space. Also, such techniques can improve the efficiency of machine learning tasks such as classification and clustering. Towards this direction, we study the behavior of seven well-known and cutting-edge dimensionality reduction techniques tailored for RNA-sequencing data. Along with the study of the effect of these algorithms, we propose the extension of the Random projection and Geodesic distance t-Stochastic Neighbor Embedding (RGt-SNE) algorithm, a recent t-Stochastic Neighbor Embedding (t-SNE) improvement. We suggest a new distance criterion for the kernel matrix construction. Our results show the potential of the proposed algorithm and, at the same time, highlight the complexity of the COVID-19 data, which are not separable, creating a significant challenge that the Machine Learning field will have to face.

Keywords: Dimensionality reduction · Single-cell RNA-sequencing · High-dimensional COVID-19 data

1 Introduction

The emergence of Coronavirus Disease 2019 (COVID-19) pandemic resulted in a great impact on global health care systems, affecting various fields and infecting millions of people worldwide [1]. The unprecedented pandemic had a great impact on scientific community, motivating researchers to construct strategies, in order to understand the inner profile and molecular structure of SARS-CoV-2 [2,3]. The immense growth of literature on COVID-19 in various topics and especially in biomedical sciences developed numerous resources in order to provide a well-established framework to reveal the underlying mechanisms of the virus and identify potential therapeutic measures [4]. The ongoing pandemic required effective solutions, and the consolidation of molecular biology and computational models is one of the first choices to deal with the crucial challenge.

The technological development in the molecular biology field is to such an extent that we can analyze a disease or a biological process in great detail at the cellular level. In recent years, many technologies have been developed which give measurements with great accuracy, offering the possibility for in-depth analysis and interpretation. All omics technologies (genomics, transcriptomics, metabolomics, etc.) have opened up new research paths to clarify a plethora of complex human diseases [10]. One of the dominant technology is transcriptomics, which through the quantification of mRNA offers gene expression information for a case under study. Part of this transcriptomics family is the very recent technology called single-cell RNA-sequencing. An emerging technology that has revolutionized molecular biology since it can analyze a plethora of cells in tissue by examining each cell individually. The critical advantage is that we have measurements with great accuracy, offering reliable and robust data [5].

Various applications of large-scale transcriptome analysis attempt to elucidate the pathophysiology of SARS-CoV-2 and determine the inner pathways of immune response [6]. RNA sequencing technique provides an essential framework that may uncover differential states of cellular response in COVID-19, while the nobility of this method raised the interest of the scientific community to create public databases that present information and single cell atlases of different experiments [9]. While advances in computational methods constantly arise, single-cell data require the appropriate practices to analyze and extract useful information through data integration and interpretation [7]. Also, the major problem with single-cell RNA-sequencing data is their complexity in terms of sample size and feature space size. A typical single-cell RNA-sequencing experiment exports expression profiles for around 20 thousand genes (feature space, dimensionality) from thousands up to hundreds of thousands of cells (sample size) [8]. Understandably, we have to deal with huge complexity where their data analysis or mining is a quite challenging task.

Machine learning is how a computer system can acquire artificial intelligence. It is called learning because it is reminiscent of how we humans learn by observing a situation. Machine learning has shown excellent performance in biomedical problems of high complexity, having offered in recent decades many promising results in the interpretation and clarification of biological mechanisms activated in complex diseases. Machine learning has already made significant progress on the coronary pandemic. As the data grows, the scope for implementing machine learning methods to lead to more substantive results for the complex pandemic problem expands. Such algorithms intend to uncover new roads to identify more in-depth mechanisms of COVID-19, creating a potential validated framework that enhances the prevention, diagnosis and treatment of the disease [11]. Therefore, the need for more machine learning tools is imperative, and the research community should turn in this direction.

The present work covers one of the above needs related to the up to 2-dimensional (2D) visualization of high-dimensional COVID-19 RNA sequencing data, contributing to a better visual interpretation by the experienced biologist or doctor. Dimensionality reduction algorithms are the dominant choice to achieve a robust 2D data visualization. In this direction, several such algorithms have been proposed in recent years tailored for RNA sequencing data. We present an extensive overview of relevant cutting-edge dimensionality reduction tools in COVID-19 case studies. A variant of the recent RGt-SNE algorithm is also proposed, which differs from the original algorithm in the construction phase of the distance table among samples (cells or tissues). This variant seems to adapt better to single-cell RNA-sequencing data, effectively addressing the 2D visualization problem for COVID-19 high-dimensional data. Also, the extensive reference and application of recent dimensionality reduction algorithms in such data underline the complexity of the task and the difficulty of finding the mechanisms that distinguish the two primary states (health vs COVID-19).

2   State-of-the-Art Dimensionality Reduction Techniques

This chapter describes five modern dimensionality reduction algorithms adapted to RNA-sequencing data. Together with the two well-known algorithms, t-SNE and UMAP, they were applied to COVID-19 scRNA-seq data, and their results are analyzed in the next chapter. For one of them (RGt-SNE), we present a variation that uncovers effectively the cell-cell similarities having more accurate outcomes in scRNA-seq data.

Generally, interpreting the dimensionality reduction process briefly and descriptively, we consider it one methodology that tries to project a set of high-dimensional vectors into a lower-dimensional space where the initial data structure and their pairwise distances remain similar to a given error [12]. Dimensionality reduction techniques aim to create a subspace that preserves as much as possible intact the usable amount of information and denoising data. The preservation of local and global structure of data provides a reliable framework to examine the proximity levels and interpret the patterns and variability of data [12]. Visualizing high-dimensional datasets through 2D plane can uncover various patterns and unleash cell heterogeneity among important aspects of an experiment [8]. Visualization of high dimensional RNA sequencing data consists a crucial step from preprocessing stages through downstream analysis, assisting researchers to perceive important insights of the underlying structure and interpret sample variations [7]. While multiple tools are proposed, t-Stochastic Neighbor Embedding (t-SNE) [16] and Uniform Manifold Approximation [15] are dominating the field, providing an efficient and novel framework [13,14]. Focusing on recently published dimensionality reduction techniques that are tailored for scRNA-seq data, we archived different contemporary state-of-the-art algorithms. Based on the flexible and novel framework of t-SNE and UMAP, the formation of variants, such as densMAP [18], den-SNE [18] and RGt-SNE [20], are examined as alternative methods for high-dimensional data visualization. Additionally, robust approaches with advanced mathematical backgrounds emerge and seem to present promising results, such as the Potential of Heat-diffusion for Affinity-based Trajectory Embedding (PHATE) [19]. Furthermore, the reliability of Self-assembling Manifold algorithm (SAM) is investigated as another alternative tool for dimensionality reduction and visualization [23]. Examining the theoretical background of these methods, various aspects and limitations can be viewed as a brief introduction to the core concepts of each technique.

More specifically, Densvis package includes the den-SNE and densMAP, two visualization algorithms for high dimensional data [18]. Highly motivated by the well-established dimensionality reduction techniques of t-SNE and UMAP especially for transcriptomics [14,17], these proposed methods present a different scope from the aforementioned techniques targeting to overcome the limitation of loss of local density of data points and overcome misleading visualizations, which represent an aggregation of clusters with low differentiation of heterogeneity dispersion that exists in reality. By introducing the notion of local radius, densMAP and den-SNE manage to apprehend the density of each data point from the original nearest neighborhood approach, aiming to take a deeper look to the insights of heterogeneity and accurate local structure [18].

The basic theoretical background of t-SNE and UMAP presents the construction of a neighborhood graph in hyper-space followed by the assignment of different probability kernels at each dimension for the edges and enhancing the quality of embedding by operating the minimization of a cost function [18]. Inheriting the fundamental steps of these noble algorithms, densMAP and den-SNE manifest prime adjustments to original probability kernel functions in high and low dimensional space. The required objective function that measures the accordance of high dimensional space and the low embedding is maximized in order for the points to be permuted in the most appropriate position. Adjustments regarding initial concepts of objectives function optimizations are incorporated as the correlation of local radius of initial and projected space is maximized. The quality of correspondence between the local radius in the original dataset and in the embedding is measured by the correlation coefficient, which represents a density preservation objective. The objective functions are augmented in a new formula, where the target is the maximization with respect to the optimizing criteria performed in both t-SNE (gradient descent) and UMAP (stochastic gradient descent) [18].

PHATE consists of a non-linear method, aiming to preserve both global and local structure [19]. It provides a much more sufficient mathematical background to distinguish different patterns between various populations. The theoretical backbone of PHATE depends on diffusion geometry while it implements some prime adjustments, providing high efficiency to visualize in two dimensions [19]. To understand the overall shape of the data, the initial step consists of the tracking of proximity measures between data points. The calculation of dissimilarity between points is achieved by searching the pairwise distances in Euclidean metric space. The conversion to local affinities requires a radial basis Gaussian kernel, which manages to quantify the spread of neighborhood [19].

Additionally, the transition probabilities between cells constitute the suitable approach to "diffuse", hence to transport from regions of high aggregation of cells to regions of low affinities via random or stochastic process in order to obtain a normalized probability matrix and capture the global structure of data (Markov transition matrix or diffusion operator) [19]. In contrast to the traditional approach of diffusion maps, instead of the instant eigendecomposition the algorithm attempts to transform the powered diffusion operator to "potential distances" by applying log-transformation to the probabilities. The resultant matrix is pipelined to the non-metric Multidimensional Scaling in order to visualize the variation of numerous data points. The "goodness of fit criterion" is modeled by setting a stress function, for which the minimization results in a much more accurate embedding [19].

The RGt-SNE methodology [20] sets its core concepts in three main sections. The reconstruction of a similarity matrix by projecting the original space to a new low dimensional embedding via Random Projection method and k nearest neighbors searching to each space. The similarity matrix is constructed after counting the number of times the sample j was the nearest neighbor of sample i and is instantly inverse to a distance matrix. The second part consists of searching for geodesic distances from each pair of samples. At last, the resulting distance matrix is pipelined to t-Stochastic Neighbor Embedding algorithm for visualization in a two-dimensional plane [20]. In contrast with the commonly-used metric of Euclidean measure, in this paper, we perform an excessive use of a distance-based metric of correlation in the local neighborhood searching, for which we observed presents better results within gene expression data, taking into account also recent scientific works that evaluated the performance of different association measures, while other recent approaches concentrate their efforts to conduct a correlation distance-based scheme for downstream analysis in gene expression networks and pathways [21,22].

Self-assembling manifold (SAM) algorithm is an ensemble method that attempts to elaborate important insights into gene expression levels [23]. By generating a random adjacency matrix via k-nearest neighbor investigation, SAM seeks patterns between different collections among expression proximity. SAM is applied through a log normalized preprocessed matrix and facilitates the Fano factor to capture variability among highly dispersed genes. The rescaling of expression matrix is implemented by multiplication of weighted graph with preprocessed log-transformed gene expression matrix, aiming to capture variance between gene expression levels. PCA is applied to the rescaled matrix, and all the PCs are used to calculate cell-to-cell redefined distances. This iterative procedure repeats until root-mean-square deviation of gene weights converge to a value close to zero [23].

Evaluation of dimensional reduction methods is applied by the use of internal clustering criteria to examine the functionality of each in approach. Distance-based measurable criteria are applied to illustrate the validity of dimensionality reduction techniques in compactness, separability, and connectedness [28]. Technically, mitigating the problem to a clustering validation weighting scheme, metrics assess the quality of the embedded structures in two-dimensional space, while hypothesizing that provided metadata, such as COVID-19 and healthy controls are the predicted clusters, even though belonging to expected categorical data. Silhouette coefficient, Dunn index, Davies-Bouldin indices and Nearest Neighbor Error (NNE) [12] are calculated in order to evaluate the performance of each approach. Silhouette information score presents the average contribution of a sample to each cluster membership, and Dunn index quantifies inter-cluster dissimilarity levels [29,30]. Both metrics are confidence indicators of compactness and how the variation in intra-cluster level is concentrated. While values approximating one are considered good for Silhouette information, Dunn index assign "well-clustered" solutions in predominantly larger values [29,30]. Davies-Bouldin metric is examined to detect the efficiency of separability and the distance between the centers of each cluster, thus how well clusters are separated from each other. In contrast to previous indices, Davies-Bouldin index approximates better clustering results for small measurements [29,30]. Nearest Neighbor Error measurements by implementing nearest neighbor classifier attempt to calculate the validity of low dimensional embedding as proposed, and validate the embedding framework, quantifying reconstruction errors and local structure quality [12].

3   Experimental Analysis

3.1   Dataset Description and Preprocessing

The reliability of five recently published dimensionality reduction algorithms and the traditional and well-known techniques of t-SNE and UMAP for 2D visualization was examined in four real COVID-19-based omics dataset from one single-cell RNA-seq and three RNA-seq experiments. In particular, the first dataset (Dataset 1- GSE152075) tries to analyze differential gene expression in SARS-CoV-2 hosts by RNA-sequencing nasopharyngeal swabs [24]. For this experiment, 430 SARS-CoV-2 positives and 54 negative controls were examined with various amounts of viral loads and different demographic characteristics such as age and sex, aiming to uncover patterns of host response to predict disease severity. Thus, the first dataset consists of 484 RNA-seq samples of, 35784 gene expressions. Due to gene counts profile of the dataset, some basic filtering of excluding columns summing to zero was performed as a preprocessing step to the dimensionality reduction algorithms, resulting to 31512 genes. The second experimental dataset (Dataset 2 - GSE163151) examines the gene expression identification of 404 patients with different clinical histories regarding disease severity and various viral pathogens that cause respiratory illnesses [25]. Specifically, among 404 individuals, 351 gene expression counts were retrieved from nasopharyngeal swabs and 53 from whole blood samples. The research attempted to investigate the crucial network activation due to immune response to different pathogens. The proposed survey used different pathogens or donor controls to elaborate significant biomarkers to construct a diagnostic approach through COVID-19 and other acute viral respiratory illnesses and sample tissue extraction. This study consisted of overall 145 SARS-CoV-2 infected patients, 175 SARS-CoV-2 negative patients with other viral infections or 82 with non-viral infections, 31 donor controls, and 5 donors with Bacterial Sepsis. In this case, we included and examined the trend of COVID-19 whole blood samples and donor controls, extracting a dataset of 176 samples with, 26486 genes (145 COVID-19 samples and 31 Donor controls). Note here that the original data contains 404 samples, however we isolated COVID-19 and Donor controls samples. Regarding the normalization step we follow the pre-process pipeline as reported by the corresponding original study.

In addition, a large-scaled multi-omics analysis (Dataset 3 - GSE157103) was obtained with 126 individuals belonging to COVID-19 and non-COVID-19 groups. The authors in the original research attempt to uncover associations between different biomolecule classes to achieve a better identification of the biological processes of the disease mechanisms and unleash potential therapeutic opportunities [26]. The final normalized dataset had 126 single-cell transcriptome samples for, 19472 genes in two clinical conditions of COVID-19 and non-COVID-19 patients. The latter dataset (Dataset 4 - E-MTAB-9221) contains peripheral blood samples of 25 control (12) and COVID-19 (13) patients examining different biomarkers and analyzing them by spectral flow cytometry, aiming to discriminate patients that require instant hospitalization [27]. Data was downloaded by ArrayExpress repository where preprocessing steps such as, normalization and scaling have already been performed, thus we obtained expressions of, 18958 genes for 6178 cells including the two classes for COVID-19 infected (4527 cells) and control cells (1651 cells) (Table 1).

Table 1. RNA sequencing datasets description

Accession	Samples	Genes	Database
GSE152075	484	31512	GEO
GSE163151	176	26486	GEO
GSE157103	126	19472	GEO
E-MTAB-9221	6178	18958	ArrayExpress

We implemented all seven dimensionality reduction methods (source code) using the default parameters as suggested by the author. Few minor configurations were made on some executions when a better visualization outcome was observed. There was no exhaustive search of parameters since in various changes no significant changes were observed in their performance. More specifically, in cases of UMAP, t-SNE and den-SNE different implementations have been conducted, while in densMAP a correlation-based metric was included in all datasets due to higher visualization quality. Additionally, RGt-SNEcorr is dependent on parameters of the number of random projection spaces, the projected dimension, and the number of nearest neighbors, while correlation-based kernels were used exclusively in all cases. Specifically, while no parameter analysis was performed in our survey, the behavior of the most sensitive parameters of each algorithm was investigated such as perplexity and number of neighbors in t-SNE and UMAP respectively. Perplexity is proposed to range between 5 and 50, thus t-SNE and den-SNE algorithms were examined among this interval [14]. The number of neighbors was also assessed between a range of 5 to 30 for UMAP and densMAP methods. In cases of den-SNE and densMAP, variations of the well-established techniques of t-SNE and UMAP, density preservation parameters remained at the predefined level of the software creators.

Perplexity was assigned to 25 for den-SNE in all datasets, while in t-SNE adjustments were made for better visualizations. Values of 25 were defined for datasets 1 and 3, while perplexity equal to 30 was preferred for datasets 2 and 4. In the case of UMAP, number of neighbors was set to 15 to all datasets, while in the density preservation method of densMAP number of neighbors followed the same tuning except Dataset 1, in which the parameter was defined as 30. PHATE and SAM followed default approaches in parameter tuning for all datasets. For example, parameters of number of nearest neighbors for kernel generation were set to 5 with the decay level kept to 40. The sensitivity of different combinations of parameters regarding optimization and the novelty of RGt-SNEcorr is not examined in the presented research. After an extensive search, we concluded that slightly better visualizations were obtained for 30 random projected spaces, in which random matrix was generated following uniform distribution as it was constructed to the original paper [20]. The number of random projected spaces reached no significant visual improvement for highly increased values, thus Dataset 1 and Dataset 2 were assigned to 20 and 10 respectively, while datasets 3 and 4 were allocated to 7. The graph of the proximity measures for each sample was constructed after the application of the nearest neighbor search. The number of nearest neighbors was set to 3 for the first, third, and fourth datasets, while for Dataset 2 we searched for the first 5 nearest neighbors. Overall, each technique was operated on a server with 3.80 GHz Intel Core i-7 CPU and 125 GiB of memory, thus the computational cost of each method did not present significant deviations as each algorithm was executed in few seconds.

3.2   Results

All dimensionality reduction methods were applied to four RNA sequencing studies. The 2D visualizations (see Fig. 1) show a slight predominance of our RGt-SNE correlation distance-based method compared to PHATE, and den-SNE, two state-of-the-art techniques tailored for scRNA-seq data. Also, the data complexity can be easily observed since the two main classes (COVID-19 and non-COVID-19) are not easily separable. The labels that were used from each dataset attempted to analyze the behavior of our data regarding the distinguishing of COVID-19 positive samples with negative viral loads or healthy controls. For the first dataset (GSE152075), RGt-SNEcorr manages to provide a

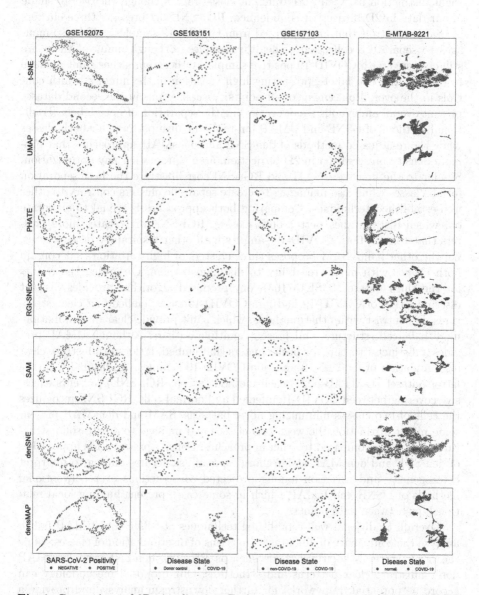

Fig. 1. Comparison of 2D visualization. Each point represents a cell (E-MTAB-9221) or sample (GSE157103, GSE163151, GSE152075) and each color represents a different COVID-19 disease state according to original data biologically meaningful annotation. Datasets complexity in terms of COVID-19 and healthy population's distinction is illustrated above. We observe the superiority of the RGt-SNEcorr and its ability to distinguish the 2 classes, which are not easily separable from the other well-established methods.

slight distinction between both different.classes. Even though the overall shape of our data in 2D projection is undefined, RGt-SNEcorr presents three clusters of healthy controls that are separated from COVID-19 samples. SAM also manages to separate a collection of healthy samples, although many of them are still clustered with COVID-19 positive samples. Other approaches can't manage to distinguish both labels, presenting high variation of the minor's health controls in the overall placement of all points. Visualizations in the second dataset present overall ambiguous results regarding the separation of both labels. Traditional methods of t-SNE and UMAP fail to distinguish patterns for both classes while the re-adjusted methods of den-SNE and densMAP, approaches that promote density preservation in 2D projection, aren't promoting any optimization. Similar to the first case, SAM and RGt-SNEcorr illustrate the best separation for all classes while also managing to create separable clusters contrary to other well-established techniques. Comparing both approaches is trivial to elaborate on, which distinguishes better both classes. RGt-SNEcorr creates more clusters that contain both COVID-19 samples with other respiratory disease cases. On the other hand, SAM proposes three clusters of high variation that contain both classes with mediocre ability to distinguish them. Observing the results for the third dataset (GSE157103), our proposed algorithm provides distinct discrimination of COVID-19 and non-COVID-19 cases, embedding clear structures, in contrast with other methods, which can't provide clear discrimination, more balanced structure, and present multiple outliers. Although UMAP provides a distinct structure for both clusters that embed, it doesn't illustrate clear discrimination of COVID-19 and non-COVID-19 samples. Additionally, in the latter dataset (E-MTAB-9221) each method (except RGt-SNE) presents excessive covering between COVID-19 infected and normal cells, RGt-SNE recognizes the cells of two classes minimizing noisy patterns. SAM and densMAP present some minor distinctions between two classes and manage to create solid structures of clusters compared to other approaches. Density preservation approaches of den-SNE and densMAP assign more "spread" structures aiming to interpret heterogeneity and variation among projected constructions over conventional methods of t-SNE and UMAP which in some cases present binding local relationships between data points.

Overall, traditional well-established techniques of t-SNE and UMAP fail to analyze the complexity of all datasets in terms of discrimination of classes, while the other state-of-the-art methods such as PHATE, den-SNE and densMAP don't illustrate clear patterns and structures. Our proposed methodology can accord a "potential framework" for further downstream analysis and provide a much more excessive interpretation due to the clearer discrimination of classes, exploiting additional metadata information of both datasets. Visualization of each dataset unleashed the complexity that is underlying in the distinct separation of COVID-19 and control samples. Arbitrary cluster structures are highlighted, observing the direction points are aligned from high dimensional to 2D embeddings.

Table 2 contains the aforementioned distance-based metrics for each visualization scheme. The visualization complexity of all datasets for each technique is

Table 2. Measurable visualization results

Method	Silhouette	Dunn Index	Davies Bouldin	NNE
Dataset 1 - GSE152075				
densMAP	**3,62E−01**	**7,36E−03**	**5,06E−01**	**6,20E−03**
den-SNE	1,06E−02	1,69E−03	9,24E+00	1,85E−01
PHATE	−1,34E−02	2,26E−04	1,35E+01	1,68E−01
RGt-SNEcorr	1,66E−01	2,58E−03	1,95E+00	8,95E−03
SAM	8,60E−02	2,01E−03	3,06E+00	9,44E−02
t-SNE	2,60E−02	6,10E−03	5,40E+00	1,45E−01
UMAP	−6,34E−02	1,26E−04	1,08E+01	1,84E−01
Dataset 2 - GSE163151				
densMAP	**6,74E−01**	8,81E−04	1,10E+00	8,71E−02
den-SNE	3,50E−01	1,31E−03	1,79E+00	1,27E−01
PHATE	5,84E−01	2,02E−04	1,58E+00	1,61E−01
RGt-SNEcorr	5,80E−01	**2,78E−02**	**7,30E−01**	**1,70E−02**
SAM	1,78E−01	2,04E−03	2,00E+00	9,66E−02
t-SNE	2,77E−01	2,05E−03	1,33E+00	1,21E−01
UMAP	4,50E−01	1,51E−03	1,10E+00	1,34E−01
Dataset 3 - GSE157103				
densMAP	1,07E−01	3,83E−03	4,01E+00	2,06E−01
den-SNE	8,83E−02	7,38E−03	3,58E+00	2,49E−01
PHATE	9,27E−02	1,08E−03	3,01E+00	2,51E−01
RGt-SNEcorr	**2,69E−01**	1,55E−03	**7,54E−01**	**1,06E−01**
SAM	7,36E−02	1,34E−02	3,53E+00	2,09E−01
t-SNE	9,30E−02	**1,46E−02**	3,13E+00	2,30E−01
UMAP	1,54E−01	2,34E−03	3,21E+00	2,30E−01
Dataset 4 - E-MTAB-9221				
densMAP	4,02E−02	2,32E−05	8,89E+00	1,97E−01
den-SNE	−9,53E−02	2,81E−05	1,85E+01	1,74E−01
PHATE	−1,40E−01	9,10E−07	5,21E+00	2,32E−01
RGt-SNEcorr	**9,55E−02**	7,55E−05	**2,40E+00**	**9,17E−03**
SAM	5,04E−02	7,32E−06	9,39E+00	1,64E−01
t-SNE	−8,49E−02	**8,81E−05**	3,18E+01	1,69E−01
UMAP	−6,62E−02	1,99E−05	6,60E+01	2,08E−01

confirmed as measurements present overall poor performance quality. In particular, PHATE performs poorly in most cases, failing to create distinct structures.
RGt-SNEcorr outperforms well-established methods, providing a more efficient
framework to interpret and manifest results. Traditional state-of-the-art manifold learning algorithms of t-SNE and UMAP display significant low-quality

structures. Comparing them to recently density preservation adjusted methods of den-SNE and densMAP, we can conceive that den-SNE doesn't surpass t-SNE. On the other hand, densMAP manage to outperform in some indices measurements and provide the best scores in the first dataset (GSE152075) and develop a reliable framework compared to UMAP. The investigation of SAM shows ambiguous results regarding the efficiency and the validity of internal scores although, manages in many cases to surpass well-established methods of t-SNE, den-SNE and PHATE. Distance-based metrics proved the low quality resulted in compactness and separability that was demonstrated through indefinable structures in various visualizations. Density preservation methods of den-SNE and densMAP tend to show higher separability than original t-SNE and UMAP. Dataset 2 (GSE163151) denotes the best measurements and visualization schemes presenting higher quality performance measurements, resulting in defined clusters and discrimination between expected label annotation.

3.3   Discussion

Considering the aforementioned results, we can clearly state that COVID-19 is a complex disease that is difficult to clarify. Our comparative study with 2D visualizations highlights the data complexity. Although we selected cutting-edge dimensionality reduction tools tailored for RNA sequencing data, the huge complexity of COVID-19 did not allow the tools to find clearly patterns that separate the health and the COVID-19 disease state. It is worth mentioning that the four datasets were obtained from different experiments and paradigms that aim to unleash inner mechanisms of COVID-19 infection, with provided metadata that presents different characteristics and insights. Observing the imbalanced labeling for each dataset, we aimed to compare algorithmic methods for a clearer distinction of different clinical conditions, an approach that may provide an important framework to unleash important insights into different populations.

As a result of large-scale analysis, high-dimensional data provide a challenging field due to the computational cost and high complexity. As the number of dimensions increase, measuring proximity between cell (or tissue) populations are less reliable, while noisy patterns distort the information and variability of data. Higher dimensionality requires high computation cost and machine learning algorithms tend to lose their efficiency due to the "Curse of Dimensionality". Transitioning through lower dimensions and providing an embedding that captures the essential information while preserving both local and global cell populations' structure raises the research interest in this field [19]. As a necessary procedure of downstream analysis, dimensionality reduction aims to retain a form of high-dimensional data as it compresses the high embedding to a lower subset. Remarkably, the projection from a high dimensional space to a 2D representation consists of an important step to illustrate and assist the community in analyzing and exploring various insights and patterns [7]. Towards this direction, several tools have been recently proposed for high throughput RNA sequencing data visualization [31]. However, new dimensionality reduction and visualization techniques will be emerging, as the need for tools that can retrieve and depict the information into a 2D scheme provides essential benefits.

4   Conclusions

We studied COVID-19 through multidimensional data from transcriptomics technologies in the present work. To best of our knowledge, this is the first extensive study which highlights the dimensionality reduction techniques impact for high-dimensional COVID-19 data. These data are characterized by solid complexity and great dimension, making it challenging to analyze and extract knowledge from them. We focused on one of the essential parts of unsupervised learning, which is data visualization through dimensionality reduction algorithms. Seven cutting-edge dimensionality reduction algorithms were extensively reported and applied in four COVID-19 transcriptomics data sets. A variant of a recent dimensional reduction algorithm was also proposed, which in this variation concerns the determination of the relationships between the cell samples to create a new search space that will give more realistic results. When the algorithms are applied, our results showed that COVID-19 disease is difficult to clarify through data. Such data has substantial complexity and huge dimensionality. The continuous advancement of technology in the biomedical field offers a wealth of heterogeneous data of large volume and large dimension, thus creating the need for more dimensional tools to clarify complex COVID-19 disease.

Acknowledgements. This project has received funding from the Hellenic Foundation for Research and Innovation(HFRI) and the General Secretariat for Research and Technology (GSRT), under grant agreement No 1901.

References

1. Ioannidis, J.P., Salholz-Hillel, M., Boyack, K.W., Baas, J.: The rapid, massive growth of COVID-19 authors in the scientific literature. R. Soc. Open Sci. **8**(9), 210389 (2021)
2. Bohn, M.K., Hall, A., Sepiashvili, L., Jung, B., Steele, S., Adeli, K.: Pathophysiology of COVID-19: mechanisms underlying disease severity and progression. Physiology **35**(5), 288–301 (2020)
3. Feng, W., et al.: Molecular diagnosis of COVID-19: challenges and research needs. Anal. Chem. **92**(15), 10196–10209 (2020)
4. Qi, C., et al.: SCovid: single-cell atlases for exposing molecular characteristics of COVID-19 across 10 human tissues. Nucleic Acids Res. **50**(D1), D867–D874 (2022)
5. Saliba, A.E., Westermann, A.J., Gorski, S.A., Vogel, J.: Single-cell RNA-seq: advances and future challenges. Nucleic Acids Res. **42**(14), 8845–8860 (2014)
6. Wilk, A.J., et al.: A single-cell atlas of the peripheral immune response in patients with severe COVID-19. Nat. Med. **26**(7), 1070–1076 (2020)
7. Luecken, M.D., Theis, F.J.: Current best practices in single-cell RNA-seq analysis: a tutorial. Mol. Syst. Biol. **15**(6), e8746 (2019)

8. Sun, S., Zhu, J., Ma, Y., Zhou, X.: Accuracy, robustness and scalability of dimensionality reduction methods for single-cell RNA-seq analysis. Genome Biol. **20**(1), 1–21 (2019)

9. Fernandes, J.D., et al.: The UCSC SARS-CoV-2 genome browser. Nat. Genet. **52**(10), 991–998 (2020)

10. Hasin, Y., Seldin, M., Lusis, A.: Multi-omics approaches to disease. Genome Biol. **18**(1), 1–15 (2017)

11. Abd-Alrazaq, A., et al.: Artificial intelligence in the fight against COVID-19: scoping review. J. Med. Internet Res. **22**(12), e20756 (2020)

12. Van Der Maaten, L., Postma, E., Van den Herik, J.: Dimensionality reduction: a comparative. J. Mach. Learn. Res. **10**(66–71), 13 (2009)

13. Jolliffe, I.T., Cadima, J.: Principal component analysis: a review and recent developments. Philos. Trans. R. Soc. A: Math. Phys. Eng. Sci. **374**(2065), 20150202 (2016)

14. Kobak, D., Berens, P.: The art of using t-SNE for single-cell transcriptomics. Nat. Commun. **10**(1), 1–14 (2019)

15. McInnes, L., Healy, J., Melville, J.: UMAP: uniform manifold approximation and projection for dimension reduction. arXiv preprint arXiv:1802.03426 (2018)

16. Van der Maaten, L., Hinton, G.: Visualizing data using t-SNE. J. Mach. Learn. Res. **9**(11) (2008)

17. Becht, E., et al.: Dimensionality reduction for visualizing single-cell data using UMAP. Nat. Biotechnol. **37**(1), 38–44 (2019)

18. Narayan, A., Berger, B., Cho, H.: Assessing single-cell transcriptomic variability through density-preserving data visualization. Nat. Biotechnol. **39**(6), 765–774 (2021)

19. Moon, K.R., et al.: Visualizing structure and transitions in high-dimensional biological data. Nat. Biotechnol. **37**(12), 1482–1492 (2019)

20. Vrahatis, A.G., Tasoulis, S.K., Dimitrakopoulos, G.N., Plagianakos, V.P.: Visualizing high-dimensional single-cell RNA-seq data via random projections and geodesic distances. In: 2019 IEEE Conference on Computational Intelligence in Bioinformatics and Computational Biology (CIBCB), pp. 1–6. IEEE (2019)

21. Pardo-Diaz, J., Bozhilova, L.V., Beguerisse-Díaz, M., Poole, P.S., Deane, C.M., Reinert, G.: Robust gene coexpression networks using signed distance correlation. Bioinformatics **37**(14), 1982–1989 (2021)

22. Liesecke, F., et al.: Ranking genome-wide correlation measurements improves microarray and RNA-seq based global and targeted co-expression networks. Sci. Rep. **8**(1), 1–16 (2018)

23. Tarashansky, A.J., Xue, Y., Li, P., Quake, S.R., Wang, B.: Self-assembling manifolds in single-cell RNA sequencing data. Elife **8**, e48994 (2019)

24. Lieberman, N.A., et al.: In vivo antiviral host transcriptional response to SARS-CoV-2 by viral load, sex, and age. PLoS Biol. **18**(9), e3000849 (2020)

25. Ng, D.L., et al.: A diagnostic host response biosignature for COVID-19 from RNA profiling of nasal swabs and blood. Sci. Adv. **7**(6), eabe5984 (2021)

26. Overmyer, K.A., et al.: Large-scale multi-omic analysis of COVID-19 severity. Cell Syst. **12**(1), 23–40 (2021)

27. Silvin, A., et al.: Elevated calprotectin and abnormal myeloid cell subsets discriminate severe from mild COVID-19. Cell **182**(6), 1401–1418 (2020)

28. Handl, J., Knowles, J., Kell, D.B.: Computational cluster validation in postgenomic data analysis. Bioinformatics **21**(15), 3201–3212 (2005)

29. Rendón, E., Abundez, I., Arizmendi, A., Quiroz, E.M.: Internal versus external cluster validation indexes. Int. J. Comput. Commun. **5**(1), 27–34 (2011)

30. Bolshakova, N., Azuaje, F.: Cluster validation techniques for genome expression data. Signal Process. **83**(4), 825–833 (2003)
31. Cakir, B., Prete, M., Huang, N., Van Dongen, S., Pir, P., Kiselev, V.Y.: Comparison of visualization tools for single-cell RNAseq data. NAR Genomics Bioinform. **2**(3), lqaa052 (2020)

Soft Brain Ageing Indicators Based on Light-Weight LeNet-Like Neural Networks and Localized 2D Brain Age Biomarkers

Francesco Bardozzo[1], Mattia Delli Priscoli[1], Andrea Gerardo Russo[2,3],
Davide Crescenzi[1], Ugo Di Benedetto[1], Fabrizio Esposito[3],
and Roberto Tagliaferri[1(✉)]

[1] Neurone Lab - DISA-MIS, Università degli Studi di Salerno, Fisciano, Italy
robtag@unisa.it
[2] Dipartimento di Medicina, Chirurgia e Odontoiatria "Scuola Medica Salernitana",
Università degli Studi di Salerno, Baronissi, Italy
[3] Dipartimento di Scienze Mediche e Chirurgiche Avanzate, Università degli Studi
della Campania, Napoli, Italy

Abstract. In recent years, there have been several proposed applications based on Convolutional Neural Networks (CNN) to neuroimaging data analysis and explanation. Traditional pipelines require several processing steps for feature extraction and ageing biomarker detection. However, modern deep learning strategies based on transfer learning and gradient-based explanations (e.g., Grad-Cam++) can provide a more powerful and reliable framework for automatic feature mapping, further identifying 3D ageing biomarkers. Despite the existence of several 3D CNN methods, we show that a LeNet-like 2D-CNN model trained on T1-weighted MRI images can be used to predict brain biological age in a classification task and, by transfer learning, in a regression task. In addition, automatic averaging and aligning of 2D-CNN gradient-based images is applied and shown to improve its biological meaning. The proposed model predicts soft biological brain ageing indicators with a six-class-balanced accuracy of ≈70% by using the anagraphic age of 1100 healthy subjects in comparison to their brain scans.

Keywords: Neuroimaging · Brain age prediction · 2D brain age biomarkers · Brain age biomarker explanations · Brain ageing detector

1 Introduction

Anagraphic age differs from biological brain age (BBA) [1]. Through the analysis of T1-weighted magnetic resonance imaging it is possible to study BBA biomarkers to foreshadow patient neurodegenerative diseases [2,3]. Traditional machine learning algorithms and 3D based convolutional neural networks (3D-CNN) are applied to identify ageing biomarkers. In turn, 3D-CNNs could provide a plethora of information even if they are both time-consuming in preprocessing

© The Author(s), under exclusive license to Springer Nature Switzerland AG 2022
D. Chicco et al. (Eds.): CIBB 2021, LNBI 13483, pp. 242–252, 2022.
https://doi.org/10.1007/978-3-031-20837-9_19

phases and require high computational resources in training phases [4,5]. On the counterpart, traditional methods may be faster even if their results are less interpretable from the neurologist point of view [6]. In this work, we show that it is possible to predict the BBA by adopting a lighter and explainable approach which consists of using 2D brain slices for ad-hoc 2D neural networks with gradient-based methods for 2D aging biomarker visualizations [7]. In fact, 2D slices of the brain MRI can be analyzed to ensure classification/identification interpretability [8]. Convolutional Neural Networks (CNNs) are often used in conjunction with other techniques in order to explain why the network makes a particular prediction. These techniques are typically used to extract visual features from images. However, machine learning (ML) and deep learning (DL) models can also be used for eXplainable AI (X-AI) systems. This would allow for the behavior of the "black-box" models to be explained and their predictions interpreted [9]. There are two types of X-AI methods: those that can be implemented intrinsically, meaning that the models are already transparent and do not require further explanations (for example, Principal Component Analysis, Logistic regression, or Decision Trees); and those that can be applied after the model prediction to explain 'why' this decision was taken. The latter are largely explored in healthcare and in biomedical image analysis, ensuring the explainability of the developed models and supporting the medical scientific community in the diagnosis of diseases [10]. In [11] a Principal component analysis is applied to the CNN classification predictions of two neuroblastoma cells line in digital holograms video stream to understand why the presented ML and DL models are able or not to correctly distinguish the considered cells. In [12] a survey of the state-of-the-art X-AI methods applied to medical support decision systems are presented and categorized with respect to their type of interpretability. As a last example, in [13] a post-hoc X-AI method is applied to a DNN trained on tabular morphological features extracted from T1-weighted MRI images. Based on these works, we presented a CNN able to correctly classify 2D biological brain age indicators by reaching an average class-balanced accuracy of ≈ 0.74 on the validation set and ≈ 0.71 on the test set over a collection of 1100 subjects. Furthermore, Grad CAM++ [7] is applied as a post-hoc X-AI method to highlight the most important areas used by the CNN to correctly classify the BBA. The whole methodology is introduced in Sect. 2, while our results are shown in Sect. 3. In Sect. 4 our results are discussed. In Sect. 6, there are some conclusions about our preliminary study and future prospects.

2   Methods

The presented methodology, including the LeNet-like CNN regression and classification models and the GradCAM++ biomarkers relevance maps, is implemented using Keras [14] on top of Tensorflow [15]. The training and the inference steps were performed on an NVIDIA GeForce GTX 970 GPU with a 4 GB of VRAM. The low-budget requirements make this procedure well suited to be implemented on several embedded medical-imaging devices. In Sect. 2.1, T1-weighted MRI dataset collection, preprocessing and labelling are described

Table 1. Dataset design and distributions of 10-years-range class T1-weighted slices.

	20–30	30–40	40–50	50–60	60–70	70–80
	(a) Statistical distributions training set					
Mean age	25.4	34.83	45.15	55.37	64.61	74.57
Std age	3.22	2.66	2.91	2.91	2.88	3.03
Total subjects	124	124	124	124	124	108
Male subjects	62	62	62	62	62	46
Female subjects	62	62	62	62	62	62
	(b) Statistical distributions validation set					
Mean age	25.9	34.97	44.29	55.2	64.84	74
Std age	2.8	2.66	2.56	2.67	3.08	2.85
Total subjects	32	38	39	37	40	23
Male subjects	12	20	19	17	20	10
Female subjects	20	18	20	20	20	13
	(c) Statistical distributions test set					
Mean age	25.96	35.09	44.43	56.15	65.25	75.68
Std age	2.12	3.15	3.15	2.7	2.95	3.3
Total subjects	20	30	35	35	58	29
Male subjects	4	26	11	9	20	10
Female subjects	16	4	24	26	38	19

as the preparatory steps for Le-Net like 2D-CNN models. In Sect. 2.2, the 2D-CNN models for classification and regression tasks are described. Finally, feature visualization to explain trained networks classification is illustrated in Sect. 2.3.

2.1   Data Extraction, Preprocessing and Labeling

A well-balanced dataset of 1100 healthy subjects is obtained by combining two MRI datasets: **(i)** T1-weighted MRI public available datasets (*IXI* dataset[1] and the **(ii)** *CamCAN dataset*[2] provided by the Cambridge Centre for Ageing and Neuroscience) [16,17]. The MRIs are acquired with 3T and 1.5T systems in **(i)**, while a 3T acquisition system in **(ii)** is used. Approximately 10 min is required to preprocess 2D brain slices, which includes preliminary corrections for intensity bias and noise, segmentation of brain tissue from skull and non-brain area, and normalization to the MNI Average Brain (305 MRI) template. The training is supervised as the 2D subject slices are labeled by anagraphic 10-year ranges from 20 to 80 years. The between-classes 10-year range grouping is maintained unbiased with respect to class numerosity and sex distribution.

The training set is composed of 728 subjects, while the validation set consists of 209 subjects and, the test set is built with 207 subjects (see also Table 1). For each subject, a reduced number of slices (30 slices from 70 to 100 over $258 \times 258 \times 258$ voxels) is considered. According to neurologists, this slab, covering the most

---

[1] https://brain-development.org/ixi-dataset/.

[2] http://www.mrc-cbu.cam.ac.uk/camcan/.

of the cortical and white matter and ventricles, shows more typical 3D ageing biomarkers [18–21]. The class overlaps are avoided to preserve set independence. Meanwhile, in the regression task (used to strengthen the explainability of our results) the anagraphic age is used as a label.

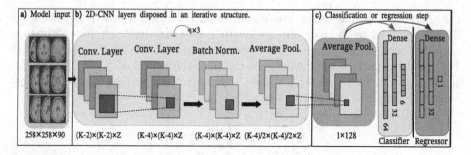

Fig. 1. The Figure is separated into three boxes. In the first box **(a)**, the input images with their respective size are shown. In the second box **(b)**, the LeNet-like convolutional neural network (CNN) architecture is shown. The output size of each layer is reported below the figure. In detail, the K and Z parameters represent the kernel size and the number of convolutional channels, respectively. Meanwhile, in the last box **(c)**, the different model endings (from blue and red arrows) show the ability for the model to be switched between a regression or a classification task. (Color figure online)

2.2   2D-CNN Models for Brain Age Classification and Regression

The LeNet-like [22] architecture is used for the 2D brain ageing classification and regression tasks. The number of neurons in the last dense layer is variable, with six neurons for classification (according to the six 10-years-range classes) and one neuron for the regression-oriented model (single subject anagraphic brain age). The architecture is designed to be used for both the classification and the regression oriented model, as shown in Fig. 1. Basically, the architecture is modulated with respect to the following parameters: **(i)** Kernel size K and **(ii)** Number of convolutional channels Z. In the classification oriented model, the Categorical Cross-Entropy is used as loss to reduce the classification error with respect to the six class 10-years-range targets over a batch-size of 90 slices. The loss is minimized with a Rectified Adam optimizer [23] to prevent model over-fitting on the weight back-propagation. The Neural Network is configured with the following parameters: learning rate of 0.001 and weights decay of 0.01. The model is driven to convergence with an early stop to the $150th$ epoch. One of the 6 class 10-years-range biological age prediction (20–30, 30–40, etc.), for each subject, is decided with a majority voting schema over the 30 slices prediction pool. The regression oriented model is a modification of the previous model. In the training phase, the weights of the previous model are frozen and the model is trained only in the tail (the last dense layers as shown in Fig. 1 - Box

(c) - Regression block). The loss considered in this case is the Mean Squared Error. For each subject, the median value is estimated over 30 slices to maintain prediction age consistency with respect to the previous model.

2.3   2D Brain-Age Biomarkers Model Explanation

2D brain-age biomaker explanations come trough a region-based visualization for 2D-CNN model predictions obtained with a gradient-weighted class activation mapping tool. Grad Cam++ [24] is used to discover the most important 2D biological biomarkers by flowing into the final convolutional layer through the network graph. Thus, the Grad CAM++ logic can be described for the 10-years-range age classes c as follows:

$$M^c = ReLU \left(\sum_k \alpha_c^k A^k \right) \tag{1}$$

where:

$$\alpha_c^k = \frac{1}{N} \sum_{i,j} \frac{dy^c}{dA_{i,j}^k} \tag{2}$$

where N is the number of pixels in the input brain slices, A^k is the feature map of interest (for example the last convolutional layer in the CNN), and y^c corresponds to a scalar class score. In this way, the feature map M^c is computed to understand why the CNN is focusing on a particular area of the input brain slices. For each subject, a collection of 30 heatmaps is obtained. Then, these heatmaps are processed and normalized to mantain within- and between-subject registration with original images. This class activation map method is commonly used as a post-hoc X-AI for image detection/classification analysis, ensuring high reliability to understand the CNN decision making process.

Table 2. Balanced classification accuracy for each 10-years-range class and average classification accuracy (*Avg*)

	20–30	30–40	40–50	50–60	60–70	70–80	Avg
Train.	0.98	0.97	0.86	0.89	0.92	0.93	0.93
Val.	0.82	0.67	0.72	0.69	0.76	0.80	0.74
Test	0.73	0.68	0.65	0.61	0.71	0.87	0.71

3   Results

In Sect. 3.1, the performance of the classification model is described using confusion matrices that take into account the class-balanced and averaged accuracy. The class-balanced accuracy is calculated by weighting the six-class scores according to the number of subjects in each age group. For this reason, those values are higher and meaningful with respect to the averaged one. Furthermore, in Sect. 3.2, the results obtained from the linear regression model are evaluated by means of their median predictions $\widehat{m}$, their mean absolute error (MAE) [25] and Pearson's correlation coefficient (ρ^2) with respect to the true values [26].

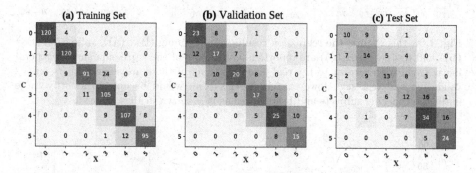

Fig. 2. In this Figure, the classification results from training (Box **(a)**), validation (Box **(b)**) and test set (Box **(c)**) are reported. For each dataset, the confusion matrix for the 6 classes of 10-year age ranges is reported. Both the age ranges and the classes are shown with 6 labels from 0 to 5.

3.1 Classification Results

The percentages of matching between the anagraphic age and the biological predicted one are the following: 87.64% on 728 subjects on the Training set, 55.98% on 209 subjects on the Validation set and 51.69% on 207 subjects for the Test set. As shown in Fig. 2, most of the predictions are on the diagonals of the confusion matrix, which indicates that the biological and the anagraphic ages are generally consistent. Following the figure, most of the misalignments are in the classes that are adjacent to the correct ones, showing interesting behaviors that are better explained with the class-balanced accuracy (Table 2). Finally, the classification power of our 2D-CNN is shown in Table 2 - *Avg* column.

3.2 Linear Regression Results

The results of the regression task can be seen in Fig. 3. In detail, the median value $\widehat{m}$ of the within-subjects predictions over 30 slices are collected (see also Sect. 2.1). In particular, on the training set, the results of the regression model show a $MAE = 4.19$ and $\rho^2 = 0.95$ of $\widehat{m}$ with respect the true value. The MAE and *Pearson's correlation coefficient* slightly change for Validation set ($MAE = 5.03$ and $\rho^2 = 0.91$). Finally, in the test set, the $MAE = 6.06$ and a

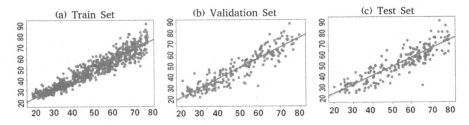

Fig. 3. In this Figure, the regression results from training (Box **(a)**), validation (Box **(b)**) and test set (Box **(c)**) are reported. The results of the regression show a general linear trend with respect to the real brain age, even if there are some outliers on the validation and test set. However, our X-AI system would be able to identify these outliers on specific subjects and brain slices as it is shown in Figs. 5 and 6.

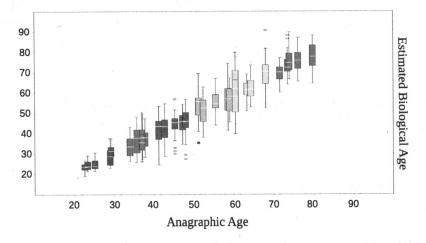

Fig. 4. The Figure shows the relationships between the predicted subjects biological age (y-axis) and their anagraphic age (x-axis) in a box-plot visualization. The different color groups in the graph represent the 10-year age range classes as described in Sect. 2.2. In particular, each box-plot represents the variability of ageing estimation on different slices of a single subject.

$\rho^2 = 0.89$. In Fig. 4, the variability of predictions within subjects and between slices is shown by box plots of different colors (one for each 10-year-range class).

4   Discussion

The 2D-CNN LeNet-like models explained with GradCam++ over 2D brain slices can effectively classify soft biological brain age indicators with a certain accuracy. As it is shown on the explanation maps of Fig. 5, in most cases, the ageing process is aligned with the anagraphic age (i.e. the age indicated on a person's birth certificate). However, there are some cases where there is an

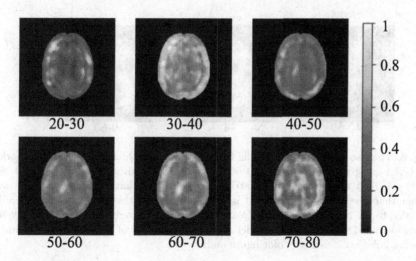

Fig. 5. The Figure shows the explanation map of every 2D brain biomarker for age classification by averaging for each age range. The maps of importance were obtained through the application of GradCam++ on every slice for each subject and then mediated.

evident misalignment between the two. In these cases, according to *Smith et al. 2019* [27], the age discrepancy or alignment between ageing and anagraphic age can be determined over minimum age differences.

In addition, observing visual explanation maps (e.g. those reported in Fig. 5) could help to better understand ageing, even when there is a discrepancy between chronological and biological brain age. For example, through the heat map analysis (i.e. slide 90 - Fig. 6), it emerges that the areas with greater intensity (the yellow ones) are those known to neuroradiologists to often show clues about the signs of brain aging [18–21] at least in older subjects.

However, many subjects who were classified as incorrect are actually expected to be associated with neighboring classes, not with their chronological age. This is because these subjects have neuromorphological features that are similar to those predicted in the neighboring classes, suggesting that the misaligned subjects have clear discrepancies between their chronological age and their biological age.

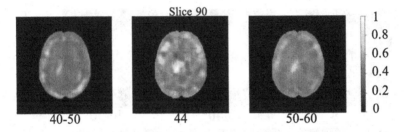

Fig. 6. The two external maps in the figure above show the most relevant biomarkers for brain aging in subjects aged 40–50 and 50–60. In addition, a single subject aged 44 is classified in the confusion matrix as an outlier falling in the 50–60 age range rather than the 40–50 age range. It is also evident that this subject is more similar, visually, to the block of averaged explanation maps of the age range 50–60. The yellow areas identified with our X-AI model are often indicative of brain aging according to neuroradiologists [18–21]. (Color figure online)

5   Architectural, Qualitative and Performance Comparisons

Our proposed LeNet-like CNN is suitable for real-time/mobile applications. In general, it shows fast performance in both the inference (about 100 ms to classify a single brain image) and in the training steps. This architecture is much lighter than others, involving only ≈120K parameters instead of the millions of parameters required by state-of-the-art 3D-CNNs trained on the same task with similar brain datasets like those of [4,28,29]. According to [30], even though the accuracy of prediction is relatively lower than 3D models in the Literature [28,29,31,32], the threshold between precision and lightness of the model can be interesting in all those systems where a preliminary investigation is necessary in real-time applications. This approach is inspired by the modeling adopted on MobileNet [33] applications. MobileNets are a class of small, low-latency, low-power models that are used for classification, detection, and other vision applications on systems with low computational capacity.

6   Conclusion

An efficient and accurate methodology for brain age classification has been proposed, based on convolutional neural networks. Unlike common 3D-CNN models adopted in the Literature, the proposed methodology is based on a light-weight LeNet-like 2D convolutional neural network. The introduced methodology takes in input the patient's anagraphic age and 2D slices obtained from T1-weighted MRI images and outputs visual explanations (heatmaps) of the regions involved in the ageing classification. The dataset is composed of 1100 brain scans and the 2D-CNN has been trained on more than 700 subjects. The experimental results obtained on the test and validation sets showed an acceptable error and a good accuracy in both classification and regression. In future works, with an improvement of localization techniques based on the gradient, it will be possible to create an automatic system for brain age classification using a pre-trained 2D-CNN with two-dimension brain slicing. Furthermore, a comparison between the state-of-the-art brain aging systems based on DL will be explored, focusing on developing intrinsically and post-hoc X-AI methods. In future findings, this system could provide a detailed description of each area of the brain involved in ageing processes with lower computational costs and in real-time contexts.

Acknowledgment. The authors would like to thank the Centre for Ageing and Neuroscience (CamCAN) to provide their data collection.

References

1. Ludwig, F.C., Smoke, M.E.: The measurement of biological age. Exp. Aging Res. **6**(6), 497–522 (1980)
2. Abbott, A.: Dementia: a problem for our age. Nature **475**(7355), S2–S4 (2011)
3. Cole, J.H., et al.: Predicting brain age with deep learning from raw imaging data results in a reliable and heritable biomarker. Neuroimage **163**, 115–124 (2017)
4. Peng, H., Gong, W., Beckmann, C.F., Vedaldi, A., Smith, S.M.: Accurate brain age prediction with lightweight deep neural networks. Med. Image Anal. **68**, 101871 (2021)
5. Priscoli, M.D., et al.: Neuroblastoma cells classification through learning approaches by direct analysis of digital holograms. IEEE J. Sel. Top. Quantum Electron. **27**(5), 1–9 (2021)
6. Bardozzo, F., et al.: Motor strength classification with machine learning approaches applied to anatomical neuroimages. In: 2020 International Joint Conference on Neural Networks (IJCNN), pp. 1–8. IEEE (2020)
7. Chattopadhyay, A., Sarkar, A., Howlader, P., Balasubramanian, V.N.: Grad-CAM++: improved visual explanations for deep convolutional networks. arXiv preprint arXiv:1710.11063 (2017)
8. Dinsdale, N.K., et al.: Learning patterns of the ageing brain in MRI using deep convolutional networks. NeuroImage **224**, 117401 (2021). https://www.sciencedirect.com/science/article/pii/S1053811920308867
9. Chittajallu, D.R., et al.: XAI-CBIR: explainable AI system for content based retrieval of video frames from minimally invasive surgery videos. In: 2019 IEEE 16th International Symposium on Biomedical Imaging (ISBI 2019), pp. 66–69 (2019)
10. Payrovnaziri, S.N., et al.: Explainable artificial intelligence models using real-world electronic health record data: a systematic scoping review. J. Am. Med. Inf. Assoc. JAMIA **27**(7), 1173–1185 (2020)
11. Delli Priscoli, M., et al.: Neuroblastoma cells classification through learning approaches by direct analysis of digital holograms. IEEE J. Sel. Top. Quantum Electron. **27**(5), 1–9 (2021)
12. Tjoa, E., Guan, C.: A survey on explainable artificial intelligence (XAI): toward medical XAI. IEEE Trans. Neural Netw. Learn. Syst. **32**, 4793–4813 (2020)
13. Lombardi, A., et al.: Explainable deep learning for personalized age prediction with brain morphology. Front. Neurosci. **15** (2021). https://www.frontiersin.org/article/10.3389/fnins.2021.674055

14. Chollet, F., et al.: Keras (2015). https://github.com/fchollet/keras
15. Abadi, M., et al.: TensorFlow: large-scale machine learning on heterogeneous systems (2015). Software available from tensorflow.org. https://www.tensorflow.org/
16. Taylor, J.R., et al.: The Cambridge centre for ageing and neuroscience (Cam-CAN) data repository: structural and functional MRI, MEG, and cognitive data from a cross-sectional adult lifespan sample. Neuroimage **144**, 262–269 (2017)
17. Shafto, M.A., et al.: The Cambridge centre for ageing and neuroscience (Cam-CAN) study protocol: a cross-sectional, lifespan, multidisciplinary examination of healthy cognitive ageing. BMC Neurol. **14**(1), 204 (2014)
18. Drayer, B.P.: Imaging of the aging brain. Part I. Normal findings. Radiology **166**(3), 785–796 (1988)
19. Allen, J.S., Bruss, J., Brown, C.K., Damasio, H.: Normal neuroanatomical variation due to age: the major lobes and a parcellation of the temporal region. Neurobiol. Aging **26**(9), 1245–1260 (2005)
20. Fjell, A.M., et al.: One-year brain atrophy evident in healthy aging. J. Neurosci. **29**(48), 15 223–15 231 (2009)
21. Walhovd, K.B., et al.: Consistent neuroanatomical age-related volume differences across multiple samples. Neurobiol. Aging **32**(5), 916–932 (2011)
22. LeCun, Y., et al.: Backpropagation applied to handwritten zip code recognition. Neural Comput. **1**(4), 541–551 (1989)
23. Liu, L., et al.: On the variance of the adaptive learning rate and beyond. arXiv preprint arXiv:1908.03265 (2019)
24. Selvaraju, R.R., Cogswell, M., Das, A., Vedantam, R., Parikh, D., Batra, D.: Grad-CAM: visual explanations from deep networks via gradient-based localization. In: Proceedings of the IEEE International Conference on Computer Vision, pp. 618–626 (2017)
25. Willmott, C.J., Matsuura, K.: Advantages of the mean absolute error (MAE) over the root mean square error (RMSE) in assessing average model performance. Climate Res. **30**(1), 79–82 (2005)
26. Freedman, D., Pisani, R., Purves, R.: Statistics (international student edition). Pisani, R. Purves, 4th edn. WW Norton & Company, New York (2007)
27. Smith, S.M., Vidaurre, D., Alfaro-Almagro, F., Nichols, T.E., Miller, K.L.: Estimation of brain age delta from brain imaging. Neuroimage **200**, 528–539 (2019)
28. Leonardsen, E.H., et al.: Deep neural networks learn general and clinically relevant representations of the ageing brain. Neuroimage **256**, 119210 (2022)
29. Pardakhti, N., Sajedi, H.: Brain age estimation based on 3D MRI images using 3D convolutional neural network. Multimed. Tools Appl. **79**(33), 25 051–25 065 (2020)
30. Shabanian, M., Wenzel, M., DeVincenzo, J.P.: Infant brain age classification: 2D CNN outperforms 3D CNN in small dataset. arXiv preprint arXiv:2112.13811 (2021)
31. Ueda, M., et al.: An age estimation method using 3D-CNN from brain MRI images. In: IEEE 16th International Symposium on Biomedical Imaging (ISBI 2019), pp. 380–383. IEEE (2019)
32. Jiang, H., et al.: Predicting brain age of healthy adults based on structural MRI parcellation using convolutional neural networks. Front. Neurol. **10**, 1346 (2020)
33. Howard, A.G., et al.: MobileNets: efficient convolutional neural networks for mobile vision applications. arXiv preprint arXiv:1704.04861 (2017)

Author Index

Printed in the United States
by Baker & Taylor Publisher Services